Cambridge Elements ⹀

Elements of Paleontology
edited by
Colin D. Sumrall
University of Tennessee

THE STRATIGRAPHIC PALEOBIOLOGY OF NONMARINE SYSTEMS

Steven M. Holland
University of Georgia

Katharine M. Loughney
University of Michigan

CAMBRIDGE
UNIVERSITY PRESS

CAMBRIDGE
UNIVERSITY PRESS

University Printing House, Cambridge CB2 8BS, United Kingdom

One Liberty Plaza, 20th Floor, New York, NY 10006, USA

477 Williamstown Road, Port Melbourne, VIC 3207, Australia

314–321, 3rd Floor, Plot 3, Splendor Forum, Jasola District Centre, New Delhi – 110025, India

79 Anson Road, #06–04/06, Singapore 079906

Cambridge University Press is part of the University of Cambridge.

It furthers the University's mission by disseminating knowledge in the pursuit of education, learning, and research at the highest international levels of excellence.

www.cambridge.org
Information on this title: www.cambridge.org/9781108794732
DOI: 10.1017/9781108881869

© Steven M. Holland and Katharine M. Loughney 2020

First published December 2020

A catalogue record for this publication is available from the British Library.

ISBN 978-1-108-79473-2 Paperback
ISSN 2517-780X (online)
ISSN 2517-7796 (print)

The Stratigraphic Paleobiology of Nonmarine Systems

Elements of Paleontology

DOI: 10.1017/9781108881869
First published online: December 2020

Steven M. Holland
Department of Geology, University of Georgia

Katharine M. Loughney
Department of Ecology and Evolutionary Biology, University of Michigan

Author for correspondence: Steven M. Holland, stratum@uga.edu

Abstract: The principles of stratigraphic paleobiology can be readily applied to the nonmarine fossil record. Consistent spatial and temporal patterns of accommodation and sedimentation in sedimentary basins are an important control on stratigraphic architecture. Temperature and precipitation covary with elevation, causing significant variation in community composition, and changes in base level cause elevation to undergo predictable changes. These principles lead to eight sets of hypotheses about the nonmarine fossil record. Three relate to long-term and cyclical patterns in the preservation of major fossil groups and their taphonomy, as well as the occurrence of fossil concentrations. The remaining hypotheses relate to the widespread occurrence of elevation-correlated gradients in community composition, long-term and cyclical trends in these communities, and the stratigraphic position of abrupt changes in community composition. Testing of these hypotheses makes the stratigraphic paleobiology of nonmarine systems a promising area of investigation.

Keywords: continental, sequence stratigraphy, vertebrates, plants, invertebrates

ISBNs: 978-1-108-79473-2 (PB)
ISSNs: 2517-780X (online), 2517–7796 (print)

Contents

1 Introduction

Stratigraphic paleobiology uses the principles of sequence and event stratigraphy to interpret patterns in the fossil record (Patzkowsky and Holland 2012). Its principles have been developed mostly for marine systems, in which the occurrence of fossils is controlled primarily by two factors (Holland 1995, 2000). The first is the distribution of species along a water-depth gradient, the primary source of ecological variation at regional scales in both modern and ancient systems. The second is sequence-stratigraphic architecture, which describes how facies change vertically and laterally (therefore, how the water-depth gradient is sampled through time), how sedimentation rates vary, and where unconformities lie.

An understanding of these two factors reveals that patterns of fossil occurrence cannot be taken at face value as straightforward records of evolution and ecological change. Moreover, stratigraphic paleobiology represents a shift away from a long tradition of viewing the fossil record as being biased to, instead, thinking of it as having a predictable and interpretable structure (Holland 2017). Recognizing this structure improves our interpretations of patterns in the fossil record and allows us to understand how ecosystems change over time (Dominici and Kowalke 2007; Holland and Patzkowsky 2007; Patzkowsky and Holland 2007; Zuschin et al. 2007; Danise et al. 2019). This recognition makes it possible to differentiate between the time of origination and first occurrence in a stratigraphic column (or the time of extinction and last occurrence; Holland and Patzkowsky 2002), and to identify true patterns of morphological change through time (Webber and Hunda 2007). It enables an understanding of the true timing and tempo of turnover (Smith et al. 2001; Holland and Patzkowsky 2015; Nawrot et al. 2018), and how to best use fossil data for calibrating phylogenies (Holland 2016). The vast majority of these studies have focused on marine systems. Far less is understood about the stratigraphic paleobiology of nonmarine settings, which differ in their stratigraphic architecture and ecological gradients.

In this Element, we develop the principles of stratigraphic paleobiology for nonmarine systems, and, based on them, we present eight broad sets of hypotheses of expected patterns in the nonmarine fossil record. Our hypotheses build on studies in modern ecology, basin analysis, sequence stratigraphy, and, especially, taphonomy (e.g., Behrensmeyer 1982, 1988; Badgley 1986; Gastaldo 1988, 1992; Rogers 1993; Badgley and Behrensmeyer 1995; Aslan and Behrensmeyer 1996; Behrensmeyer et al. 2000; Wing 2005; Rogers and Kidwell 2007). Other processes, such as climate change, biotic interchange, extinction, and so on, also create patterns in the fossil record, and we do not minimize their importance. What we want to emphasize is that the fossil record reflects not only these climatological and biological processes but also the effects of stratigraphic architecture, and that these must be separated if we are to reconstruct the history of life.

In some cases, we provide examples from the fossil record that bear on these hypotheses, although our aim is not to provide a comprehensive review. In other cases, data bearing on these hypotheses may not currently exist, and collecting data to test these hypotheses is a promising avenue of future research.

We focus primarily on fluvial and lacustrine systems because they hold the majority of the nonmarine fossil record. Insights from fossil occurrences in these systems may promote similar investigations of the more limited nonmarine fossil record from eolian, volcanic, karst, and glacial systems, all of which operate under differing sets of stratigraphic processes.

2 The Nonmarine Stratigraphic Record

Most nonmarine biotas will not be preserved because they do not occur within sedimentary basins. The vast majority (84%) of nonmarine areas today are uplands, that is, they lie outside of and provide sediment to basins (Nyberg and Howell 2015; equivalent to the "extrabasinal" of Pfefferkorn 1980). Although it is true that many of these areas do currently have young (Plio–Pleistocene) sediments, often with a fossil record, most of these would be considered doomed sediments that have little prospect of long-term preservation (Holland 2016). Only the 16% of nonmarine areas that lie within basins have any prospect of preserving a fossil record. It is a sobering thought that large swaths of North America, eastern South America, southern Africa, and much of northern Asia will leave no permanent fossil record, including the many modern biodiversity hotspots in mountainous areas and other regions that lie outside of sedimentary basins (Myers et al. 2000; Rahbek et al. 2019).

The majority (49%) of continental basins today are intracratonic. Owing to how Nyberg and Howell (2015) classified basins, many of these areas could be considered the distal extensions of foreland basins (Fig. 1), such as in Argentina, northwestern India, parts of China, and the region surrounding Kazakhstan. Extensional basins (7%; Fig. 2), passive-margin basins (10%; Fig. 3), and proximal parts of foreland basins (29%; Fig. 1) represent most of the remaining basin types. Forearc basins (1%) and strike-slip basins (4%) constitute only a minor portion of areas lying within basins. The proportions of these basin types will vary over the 400-Myr Wilson Cycle of supercontinent assembly and breakup, which principally affects the relative proportions of extensional, passive-margin, intracratonic, and foreland basins (Holland 2016). Moreover, these basins differ in their

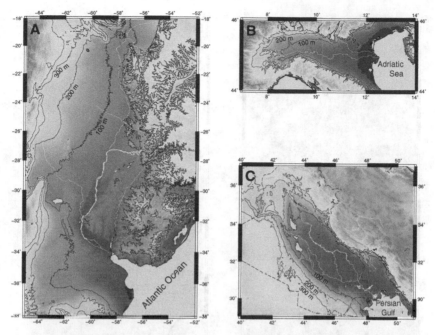

Figure 1 Relief maps of present-day foreland basins. A: Andean foreland basins of Argentina, Uruguay, Paraguay, and Bolivia. B: Po Basin of Italy. C: Persian Gulf foreland basin of Kuwait, Iraq, and Iran. Contour intervals are 100 m; contours for higher elevations outside the sedimentary basins are not shown. Map prepared with GMT (Generic Mapping Tools), using data from the National Geophysical Data Center ETOPO1 1 arc-minute global relief model (Amante 2009).

survival into the deep geologic record, with extensional, passive-margin, intracratonic, and foreland basins being more likely to survive than those associated with collisional continental margins, such as forearc basins (Sadler 2009).

These basin types differ greatly in their subsidence rates, longevity, and size (Angevine et al. 1990; Allen and Allen 2005), as well as their preservation potential (Sadler 2009). Accordingly, they differ markedly in the volume of nonmarine strata they contain and, therefore, in their contribution to the nonmarine fossil record. Foreland basins and extensional basins have fast subsidence rates, yet are relatively short-lived (Holland 2016). In contrast, passivemargin and intracratonic basins have slower subsidence rates, particularly where sediment is introduced into the basin, but they exist almost an order of magnitude longer and are far larger than foreland and rift basins (Angevine et al. 1990; Allen and Allen 2005). Finally, although forearc and backarc basins can contain substantial nonmarine records, their position along convergent margins greatly reduces their survivorship into deep time (Sadler 2009).

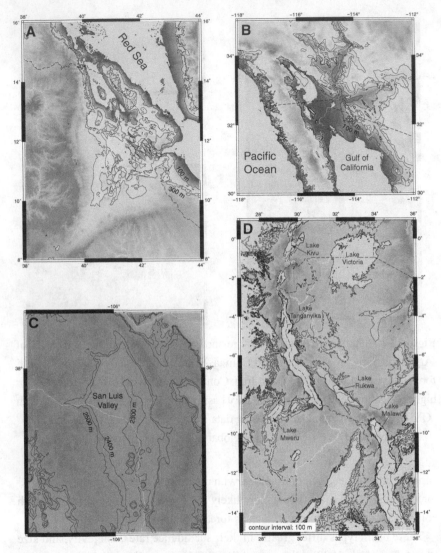

Figure 2 Relief maps of present-day coastal (A, B) and inland (C, D) extensional basins. A: Red Sea and adjacent Ethiopia. B: Gulf of California, with adjacent Mexico and California. C: San Luis Valley, part of the Rio Grande Rift System in southern Colorado, USA. D: Rift basins of East Africa. Contour intervals are 100 m; contours for higher elevations outside the sedimentary basins are not shown. Map prepared with GMT, using data from the National Geophysical Data Center ETOPO1 1 arc-minute global relief model (Amante 2009).

To understand the effects of stratigraphic architecture on the nonmarine fossil record, the starting point should be basins that are most likely to survive into deep geologic time, that contain the greatest volume of nonmarine sediments

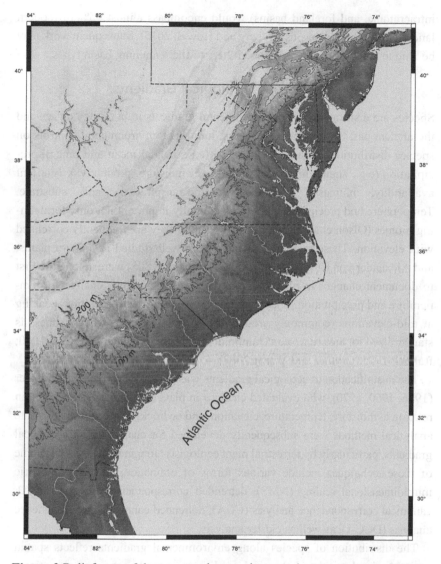

Figure 3 Relief map of the present-day passive margin on east coast of USA. Contour intervals are 100 m; contours for elevations outside the sedimentary basins (above 200 m) are not shown. Map prepared with GMT, using data from the National Geophysical Data Center ETOPO1 1 arc-minute global relief model (Amante 2009).

(foreland, including both pro-foreland/peripheral and retro-foreland/retroarc), and that are the largest and have the longest subsidence histories (passive-margin and intracratonic). Although percentages of basin types vary over geologic time, focusing on just four basin types (extensional, passive-margin, intracratonic, and foreland basins) would encompass almost 95% of modern land area lying within basins (Nyberg and Howell 2015). Subsequent work may be able to apply principles we present here to the remaining basin types.

3 Nonmarine Ecological Gradients

Species are distributed along environmental gradients in both the modern and the ancient past. In nonmarine systems, important environmental controls on species distributions include temperature (e.g., annual mean and range), precipitation (e.g., annual mean and range), soil moisture, productivity (nutrient availability, biomass), insolation, evapotranspiration, and substrate. Temperature and precipitation are fundamentally important not only for defining biomes (Olson et al. 2001), but also because they are frequently correlated with elevation. These relationships have been well studied in modern plants, and Alexander von Humboldt and Aimé Bonpland (1805) were among the first to document changes in plant communities with elevational changes in temperature and precipitation. Species richness also varies with elevation, peaking at mid-elevations (commonly around 1000–1500 m), even when richness is standardized for area (e.g., von Humboldt and Bonpland 1805; Whittaker 1960; Rahbek 1995; Grytnes and Vetaas 2002).

The quantification of ecological gradients was pioneered by Robert Whittaker (1956, 1960, 1970), who evaluated changes in plant community composition in relation to moisture, temperature, elevation, and bedrock. Numerous multivariate analytical methods were subsequently developed for characterizing ecological gradients, particularly by terrestrial plant ecologists (Jongman et al. 1995). Some of these techniques include various forms of ordination, such as nonmetric multidimensional scaling (NMS), detrended correspondence analysis (DCA), canonical correspondence analysis (CCA), detrended canonical correspondence analysis (DCCA), as well as cluster analysis.

The distribution of species along environmental gradients reflects spatial variations in physical and chemical conditions, as well as species interactions. Three types of gradients are recognized: resource, direct, and indirect gradients. Resource gradients are caused by spatial variations in the resources required by organisms, such as nutrients, food, or water (Austin et al. 1984). Direct gradients are generated by environmental variables that are not consumed by organisms but that are important controls on growth and physiology, such as moisture and temperature. Indirect gradients are formed by environmental variables that do not directly affect organisms but are linked to other environmental factors that do (e.g., water depth, elevation). Because many indirect gradients reflect the combination of several resource and direct gradients whose effects are difficult to isolate, indirect gradients are also known as complex gradients (Whittaker 1960). In this way, elevation in nonmarine systems is a proxy for a wide suite of covarying direct and resource gradients, even though elevation itself does not control the occurrence of species. Water depth in marine systems functions in the same way: water depth does not control the distribution of marine species, but it is a first-order descriptor of where species occur because it summarizes many of the factors that do control where species live.

Plants, invertebrates, and vertebrates in nonmarine systems are arrayed primarily along gradients in temperature and moisture, and these gradients are typically correlated with elevation over regional scales. Although many early studies of plant gradients (e.g., Whittaker 1956, 1960) focused on mountainous areas where elevation spanned thousands of meters, it is essential to consider the smaller elevation range spanned within sedimentary basins. Nonmarine portions of sedimentary basins have a characteristic topographic form consisting of a gently dipping plain, which may be a coastal plain in coastal basins (those that adjoin an ocean), or an alluvial plain in inland basins (those lacking a connection to the ocean). For a broad suite of basins, including those that comprise the majority of the nonmarine stratigraphic record, these alluvial and coastal plains characteristically have low relief, with a total elevation range of 200–300 m (Figs. 1–3).

3.1 Ecological Gradients on Modern Coastal and Alluvial Plains

Ecologists have widely found gradients in species composition correlated with elevation but controlled by temperature and moisture on coastal and alluvial plains, even with their modest spans of elevation. Although anthropogenic effects on modern alluvial systems can be severe (Gibling 2018), most ecological studies have attempted to use areas less impacted by human activity. Moreover, although species distributions today may be altered by human activity, it is reasonable to assume that species distributions in the past were also tied in some way to temperature and precipitation.

In coastal areas, environmental and ecological gradients are correlated with elevation and distance from the shore. For example, in the western Gulf Coastal Plain of Texas, ordination of the species composition of longleaf pine forests reveals a complex gradient reflecting the joint importance of moisture, soil texture, phosphorous, and nitrogen (Harcombe et al. 1993). On the coastal plain of North Carolina, ordination of forest vegetation demonstrates that the most important source of variation is a complex gradient reflecting moisture, physiography, stream order, and elevation (Wyant et al. 1991; Rheinhardt et al. 1998; Rheinhardt et al. 2013). Ordination of coastal plant communities in South

Africa also indicates the overriding importance of temperature and precipitation in species composition (Morgenthal et al. 2006). In inland areas, environmental and ecological gradients are correlated with elevation. The species composition of plants in the bottomland forests of the lower alluvial plain of the Mississippi River reflects flood tolerance, which is tied to floodplain slope and elevation (Dale et al. 2007). Ordination indicates that the composition of Oregon plant communities is controlled by moisture and temperature, and that these covary with elevation, a pattern that is true statewide as well as in coastal areas (Ohmann and Spies 1998). Even in arid coastal settings, such as Saudi Arabia, plant species are distributed primarily on a gradient of rainfall and elevation (Al-Aklabi et al. 2016).

Unsurprisingly, the distribution of terrestrial invertebrates and vertebrates parallels that of plants and shows similar relationships to elevation. This similarity partly reflects the way that animals are controlled by many of the same physical and chemical gradients that affect plants, but it also reflects the importance of plant–animal interactions, such as herbivory and habitat structure. For example, ordination of Scandinavian beetle communities indicates that temperature and other climatic factors are the dominant controls on their distribution (Heino and Alahuhta 2015), although this study includes areas not in sedimentary basins. Ecological diversity and species density of North American mammals are predicted by annual minimum and maximum temperature, mean annual actual evapotranspiration, relief, and elevation (Badgley and Fox 2000). These trends are most pronounced in coastal areas where the effects of increasing continentality with elevation are evident. Vertebrates of northern Australia are distributed along a gradient of rainfall and soil type, which is correlated with distance from the coast (Woinarksi et al. 1999). Some mammal species of Western Australia have distributions strongly related to distance from the coast, with others occurring more broadly (Gibson and McKenzie 2009). A perusal of field guides of many terrestrial animal groups on any coastal plain reveals a similar pattern in which some species have ranges tightly restricted to the coast, the lower coastal plain, or the upper coastal plain, often reflecting the position of the water table. Some species are more eurytopic, with a wider distribution among environments.

Possibly more surprising, aquatic animals of coastal plains are also distributed along similar gradients correlated with elevation and distance from the coast. For these aquatic species, the gradients are controlled by the upstream increase in stream gradient. For example, aquatic macroinvertebrates of the Virginia coastal plain are distributed primarily along a gradient from coastal, tidally influenced areas to nontidal, inland areas (Dail et al. 2013). This pattern is also true for macroinvertebrates on the coastal plain of North and South Carolina (Maxted et al. 2000). Unionid bivalves in the Flint River of Georgia are distributed primarily on a gradient from slack-water streams with fine sediment bottoms to shaded riffles with coarser sediment, corresponding with decreasing stream order and distance from the coast (Gagnon et al. 2006). The composition of fish communities in the Savannah River (Meffe and Sheldon 1988) and other streams of South Carolina (Paller 1994) varies along a gradient from slower-flowing, high-order streams closer to the coast to faster-flowing, low-order streams farther from the coast. Similar spatial variations are also seen in fish from eastern Costa Rica, where salinity plays a strong role in coastal areas (Winemiller and Leslie 1992).

Coastal gradients in physical and chemical properties and their effects on the distribution of species are well shown in the southeastern USA, where a wide range of data are available (Fig. 4). On a visible satellite image, the fall line that separates coastal-plain strata from upland (that is, outside of the sedimentary basin) rocks (Fig. 4, C) is marked by an abrupt change in vegetation (Fig. 4, A, G, H). Complex patterns in vegetation are also present in upland or extrabasinal areas, but as these areas have no prospect of leaving a long-term stratigraphic record, this discussion focuses only on patterns in the coastal plain. The gentle decrease in elevation towards the coast is apparent on a relief map (Figs. 3, 4, B). Although the north–south variation in ecosystems (Fig. 4, G) and bioclimates (Fig. 4, F) is partly due to latitudinal climate variation, the orientation of these ecosystem and bioclimate boundaries parallel to the Atlantic coast demonstrates the influence of the coastal-plain elevation gradient. The sources of these ecological differences are gradients in rainfall (Fig. 4, D) and temperature (Fig. 4, E) that are correlated with elevation on the coastal plain. In particular, areas closer to the coast tend to receive more rainfall than areas farther from the coast. Maximum summertime temperatures as well as mean annual temperature variations also tend to be higher inland than on the coast, owing to the buffering effects provided by the ocean. Variations in plant communities and moisture are reflected in distributions of animals. For example, birds show greater diversity in coastal areas (Fig. 4, I). At a finer scale (Fig. 4, J–L), the types of wetland habitats also vary with elevation and distance from the coast, with estuarine–marine wetlands limited to tidal areas and freshwater emergent areas just inland of the coast. Other sources of variation in community composition are also present. For example, even within areas of similar elevation, ecosystems and forest types vary with microclimate, soil type, bedrock, and human land use (Fig. 4; Edwards et al. 2013).

Elevation and distance from the coast are complex gradients (Whittaker 1956): they describe the distribution of species, but they do not directly

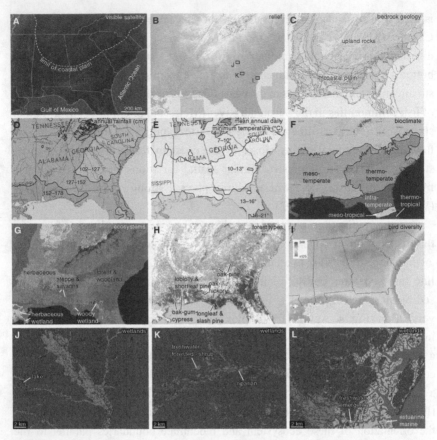

Figure 4 Elevation-related gradients on the modern coastal plain of the
southeastern USA. Locations of detailed maps J, K, and L are indicated on
B. Image sources: A–G: U.S. National Map; H: U.S. Forest Service, Forest
Types; I: BiodiversityMapping.org; J–L: U.S. Fish and Wildlife Service,
Wetlands.

control the distribution of species. These complex gradients are controlled
by direct and resource gradients that affect species distribution: these are
predominantly temperature and precipitation, but also soil type, soil moisture, stream gradient, and for animals, food sources. Because of the
covariation of environmental factors with a complex gradient, it is often
difficult to identify which environmental factor or factors are the most
important. This is true in the Recent, and even more so in ancient systems.
Even so, the distribution of species along a complex gradient can be
described and used, even if the specific underlying causes for that distribution remain unknown.

3.2 Ecological Gradients on Ancient Coastal and Alluvial Plains

Although ecological gradients in ancient nonmarine systems are seldom documented with multivariate approaches such as ordination, the studies that have been done also reflect the importance of moisture and temperature. For example, Pennsylvanian forests of the eastern USA display a moisture gradient from channel-adjacent settings dominated by *Sigillaria*, *Calamites*, ferns, and pteridosperms to mire settings dominated by lycopsid trees (Gastaldo 1987; DiMichele and Phillips 1994; Calder et al. 2006; DiMichele et al. 2007). In some Lower Pennsylvanian swamp deposits, this gradient is apparently absent, which was attributed to increasing habitat specialization during the Pennsylvanian or adaptations to changing climate conditions (Gastaldo et al. 2004); it may also represent methodological differences. Ordination of Eocene plants from Washington shows that axis 1 separates lacustrine and nonlacustrine settings, and axis 2 separates moist flood-basin settings from drier channel-margin settings (Burnham 1988). Plant community composition in the Upper Triassic Chinle Formation in Arizona varies from bennettitite-dominated, channel-adjacent, poorly drained floodplain to *Araucarioxylon*-dominated distal floodplain to gymnosperm-dominated "upland" (higher-elevation) communities (Gottesfeld 1972). Collectively, these studies indicate that environmental gradients can be preserved in the fossil record, and they may be more widely preserved than is often recognized.

Similarly, a comparison of Eocene mammals and reptiles from higher-elevation basin-margin deposits and lower-elevation basin-center deposits revealed substantial compositional differences (Gunnell and Bartels 2001). Basin-margin faunal assemblages are morphologically distinctive and more taxonomically diverse than basin-center assemblages, and these differences in species composition were attributed to greater speciation rates in the more environmentally heterogeneous higher-elevation settings. Comparisons of shore-proximal and shore-distal settings likewise support altitudinal zonation of dinosaurs (Lehman 2001; Sampson and Loewen 2005).

Quaternary studies of the eastern USA have documented individualistic responses of the distribution of plant species in response to climate change, even including the formation of no-analog communities (Jackson and Overpeck 2000; Jackson and Williams 2004; Williams and Jackson 2007). Even so, it is important to recognize these changes happened during the profound and rapid climate changes associated with continental glaciation. Moreover, some floral associations remained intact through the Quaternary despite the severity and rapidity of climate change. An important avenue of research is understanding the extent to which ecological gradients are stable through time at the scale of sedimentary basins. That communities from topographically low and high areas can be recognized in the Pennsylvanian offers promise that ecological gradients can remain intact through repeated climatic changes.

3.3 The Importance of Elevation Gradients

Although we will refer to elevation gradients throughout this Element, we want to emphasize that this is a shorthand for gradients in climate variables (temperature and precipitation), soil moisture, stream grade, and so on, expressed along a transect of increasing elevation, and in coastal basins, with distance from the shore. We refer to this as an elevation gradient to distinguish this spatial climatic control from climate change through time as well as the spatial climate variation along lines of equal elevation.

If elevation changes in the nonmarine stratigraphic record, the species composition of fossil assemblages must also change, *even if the overall climate, ecosystem, and species pool is static.* As a result, stratigraphic patterns of fossil distributions would need to be interpreted in light of the changing sampling along environmental gradients. For example, a stratigraphic change in floras or faunas may not indicate origination or extinction, but instead a change in the portion of a coastal or elevational gradient that is preserved locally (Gastaldo 1987; Looy and Hotton 2014). Additionally, changes in taxonomic diversity or composition may reflect differing rates of speciation in different parts of the landscape (Gunnell and Bartels 2001; Badgley et al. 2017). Because overall climate, ecosystem structure, and species pool may well be changing at any time, the challenge of interpreting the nonmarine fossil record is differentiating these changes from those that are caused by changes in elevation.

4 Expression of Nonmarine Ecological Gradients in the Stratigraphic Record

In light of the ubiquitous elevation-correlated gradients in nonmarine ecosystems, the topographic gradient observed in nonmarine basins becomes important to the preservation of nonmarine biotas. As topography changes over time in a basin, owing to changes in base level and sedimentation, the spatial distribution of ecological gradients will also change. For example, as a basin fills with sediment, the elevation at any given point will increase over time, as will the regional water table. Similarly, falling sea level results in an increase in elevation in coastal basins: areas that were lower coastal plain will become upper coastal plain. A simple geometric model (Fig. 5) allows elevation changes to be understood more fully.

The starting point of the model is the classic concave-up fluvial equilibrium profile, which is anchored at its upstream end (the fall line) where the stream

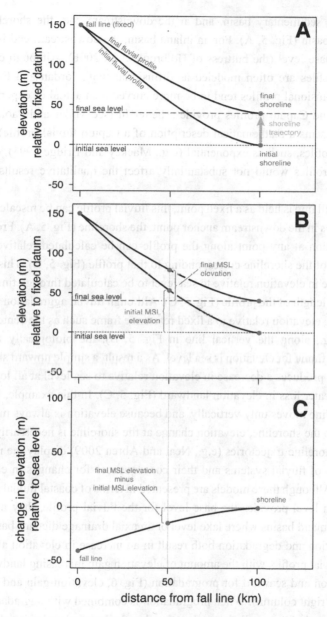

Figure 5. Response of a fluvial profile to a shoreline trajectory. A: Plot showing position of shoreline and corresponding fluvial profile at the time of initial and final sea-level positions. The upstream end of the fluvial profile is anchored at the fall line at the edge of the sedimentary basin. B: Plot showing how the initial and final elevation is calculated at one point along the fluvial profile; note that elevation is always measured relative to the current sea level. C: Plot showing the change in elevation, calculated as in B, for every point along the fluvial profile in response to a change in mean sea level (MSL) and a vertical movement of the shoreline.

enters the sedimentary basin, and at the downstream end, the shoreline for a coastal basin (Fig. 5, A). For an inland basin, the downstream end is set by regional base level (the buttress of Holbrook et al. 2006). Although concave fluvial profiles are often modeled as diffusional (e.g., Jordan and Flemings 1991), diffusional profiles tend to be more curved than actual or experimental profiles, and Caputo profiles provide a better fit (see Voller and Paola 2010, which presents a mathematical description of a Caputo fluvial profile). Other curved profiles, such as exponential (e.g., Mackey and Bridge 1995), or even straight profiles would not substantially affect the qualitative results of the model.

If the fall line is held as a fixed point, this fluvial profile can be rescaled based on changes in the downstream anchor point, the shoreline (Fig. 5, A). From this, the elevation at any point along the profile can be calculated relative to the elevation of the shoreline corresponding to that profile (Fig. 5, B). This allows the change in elevation relative to sea level to be calculated through time along the entire length of the profile (Fig. 5, C). Although fluvial aggradation leads to increasing elevation relative to a fixed reference frame such as the center of the Earth (e.g., along the vertical line in Fig. 5, B), the biologically relevant reference frame for elevation is sea level. As a result, a simple upward shoreline trajectory produces a decrease in elevation relative to sea level at all locations, with a greater loss in elevation landward (Fig. 5, C). In this example, because the shoreline moves only vertically, and because elevation is always measured relative to the shoreline, elevation change at the shoreline is necessarily zero.

Five shoreline trajectories (e.g., Neal and Abreu 2009) display the range of responses of fluvial systems and their consequences for changes in elevation (Fig. 6). Although these models are presented in terms of coastal fluvial systems where sea level provides the base level for the fluvial profile, they may also apply to inland basins where lake level or an axial drainage dictates base level. Progradation and degradation both result in an increase in elevation along the entire fluvial profile, with the amount of elevation gain increasing landward for degradation and seaward for progradation (Fig. 6, elevation gain and loss are shown in right column). When progradation is combined with aggradation, the landward end of the profile loses elevation, but the seaward end gains elevation. Under pure aggradation, the entire profile loses elevation, which is also true when aggradation is combined with retrogradation.

As the shoreline trajectory changes predictably through a cycle of rise and fall in sea level (or base level in inland systems), changes in elevation will likewise follow a predictable cyclic pattern. These changes in elevation will be reflected in cyclic changes in community composition at any location. Similarly, as a basin fills with sediment over its history and begins to uplift owing to erosion

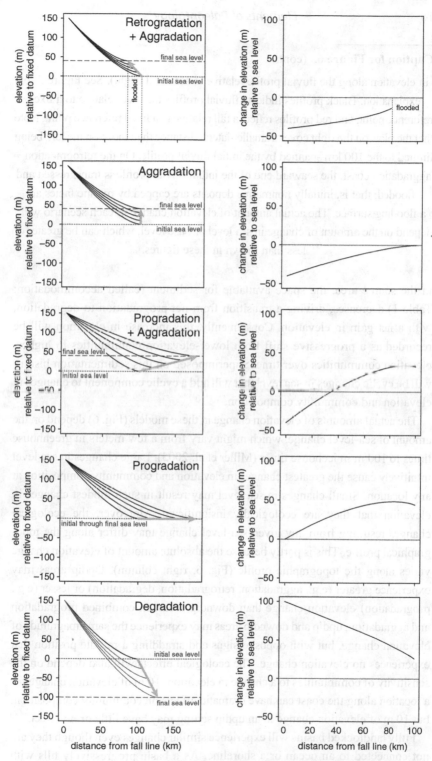

Figure 6. Response of fluvial systems to the five basic shoreline trajectories, showing the position of a fluvial profile through five time-steps (left) and the change

Caption for Figure 6. (cont.)

in elevation along the fluvial profile relative to sea level (right). See Figure 5 for
explanation. Black profiles indicate fluvial profiles that rise relative to a fixed
reference point, and red profiles reflect a fall relative to a fixed reference point. Note
that the plots on the right cover a smaller lateral distance than those on the left, being
limited to the 100 km spanned by the initial fluvial profile. On the retrogradation +
aggradation coast, the seaward end of the initial fluvial profile is transgressed and
flooded; that is, initially nonmarine deposits are capped by marine facies at
a flooding surface. The actual amount of elevation change in each scenario will
depend on the amount of change in sea level or base level, which can be greater or
less than shown in these figures.

in the source area, the space available for sediment (called accommodation;
Table 1) decreases, driving a transition from net progradation to degradation,
with a net gain in elevation. Consequently, this increase in elevation will be
recorded as a progressive shift from lower-elevation communities to higher-
elevation communities over time. Superimposed on this sedimentation history
will be cyclic changes in sea level that will add a cyclic component to changes in
elevation and community composition.

The actual amounts of elevation change in these models (Fig. 6) depend on the
amount of sea-level change, which might vary from a few meters in greenhouse
times to 100 m in icehouse times (Miller et al. 2005). Large changes in sea level
intuitively cause the greatest changes in elevation and community composition at
any location. Small changes in sea level may result in such modest changes in
elevation that they are ecologically insignificant. Moreover, the ecological
changes resulting from any given sea-level change may differ along the topo-
graphical profile. This is partly because the absolute amount of elevation change
varies along the topographic profile (Fig. 6, right column). Updip areas may
experience greater (e.g., aggradation, retrogradation, degradation) or lesser (e.g.,
progradation) elevation change than downdip areas. In combined progradation
and aggradation, updip and downdip areas may experience the same magnitude of
elevation change, but with opposite signs and straddling a middle position that
experiences no elevation change. The ecological effects will also depend on the
sensitivity of communities to a change in elevation: 10 m of elevation change for
a location along the coast can have dramatic effects on community composition,
but 10 m of elevation change in an updip setting may have little or no effect.

Fully landlocked basins will experience similar changes even though they are
not connected to an ocean or a shoreline. As a basin progressively fills with
sediment over long time scales, any location in it will experience a net increase

Table 1 Glossary of terms, with definitions based on Behrensmeyer and Kidwell (1985), Van Wagoner et al. (1988, 1990), Behrensmeyer (1991), Hunt and Tucker (1992), Martinsen et al. (1999), Catuneanu (2006), Rogers and Kidwell (2007), and Wadsworth et al. (2010).

Accommodation	Sum of the vertical rates of subsidence (or uplift) and eustatic sea-level rise (or fall). Often described as the space available for sediments to accumulate.
Aggradational stacking	A stacking pattern in which parasequences or sequences are stacked directly on top of one another, such that there is no long-term net landward or seaward drift in the position of facies belts.
Accommodation to sedimentation ratio	Ratio of the vertical rates of accommodation and sedimentation. Rates greater than 1 indicate an excess of accommodation over sedimentation; rates between 1 and 0 indicate an excess of sedimentation over accommodation; rates less than 0 indicate a combination of uplift and sea-level fall that produces a net loss of accommodation.
Base level	In fluvial systems, the lowest level toward which erosion progresses. It is approximately equal to sea level in coastal settings, and to lake level or the master axial drainage in inland fluvial systems.
Bonebed	Relative concentration of vertebrate hard parts (bones, teeth) that is derived from more than one individual and is preserved in a localized area or in a stratigraphically limited sedimentary unit.
Compound coal	A coal representing more than one genetic episode of mire formation and peat accumulation. Compound coals commonly contain one or more hiatuses and represent multiple phases of progressive wetting or drying of a mire.
Degradational stacking	A stacking pattern in which parasequences or sequences stack seawards and downwards, that is, down the depositional profile, in contrast to progradational stacking, in which units stack seawards and upwards. In nonmarine systems, degradational stacking is preserved as remnants of terraces stranded during net downcutting and valley incision.

Table 1 (cont.)

Depositional sequence	Sedimentary cycles bounded by subaerial unconformities and their correlative surfaces. In coastal scttings, a depositional sequence consists of four systems tracts defined by stacking pattern and position within a sequence. In ascending order, these are the lowstand systems tract, transgressive systems tract, highstand systems tract, and falling-stage systems tract. In inland nonmarine systems, a depositional sequence consists of a lower low-accommodation systems tract, a middle high-accommodation systems tract, and an upper low-accommodation systems tract, although this uppermost systems tract is commonly removed by erosion at the overlying sequence boundary. Within any given region, one or more systems tracts may be missing, owing to nondeposition or erosion.
Expansion surface	Relatively abrupt contact in alluvial systems between an underlying low-accommodation systems tract and an overlying high-accommodation systems tract.
Falling-stage systems tract (FSST)	The fourth and uppermost systems tract within a coastal depositional sequence, characterized by degradational stacking.
Flooding surface	Sharp contact separating overlying deeper-water facies from underlying shallow-water facies. Surface may display minor erosion, fossil accumulations, and firmground or hardground features. In nonmarine settings, coal beds often form at the updip equivalent to a marine flooding surface.
Fossil concentration	Sedimentary horizon containing abundant fossil material, distinct from overlying and underlying horizons that contain little or no fossil material.
High-accommodation systems tract (HAST)	Systems tract developed in inland nonmarine strata, containing extensive floodplain mudstone deposits with isolated single-story channel sandstones.
Highstand systems tract (HST)	The third systems tract within a coastal depositional sequence, characterized by aggradational to progradational stacking.

Table 1 (cont.)

Hydromorphic paleosol	Ancient soil formed in association with a consistently elevated water table or impeded surface-water drainage, which slows the breakdown of organic matter and promotes the reduction of iron and manganese. Typically identified by gray, blue, or green (low chroma) color. Also called a poorly drained paleosol.
Low-accommodation systems tract (LAST)	Systems tract developed in inland nonmarine strata, containing extensive amalgamated multistory channel sandstones and limited floodplain mudstone deposits.
Lowstand systems tract (LST)	The lowest systems tract within a coastal depositional sequence, characterized by progradational to aggradational stacking.
Mature paleosol	Ancient soil with well-developed pedogenic features such as horizonation, blocky peds, root casts or traces, redoximorphic features, or nodules, reflecting a prolonged period of exposure and soil development.
Mire	Encompassing term for all peat-forming environments, including swamps, marshes, and bogs
Ombrotrophic	A mire in which the groundwater is supplied primary from rainfall rather than subsurface flow. As a result, the groundwater table can be locally perched above the regional groundwater table. Also called raised mires.
Parasequence	Sedimentary cycles bounded by flooding surfaces. Internally, parasequences typically have simple, shallowing-upward arrangements of facies bound through Walther's Law.
Progradational stacking	A stacking pattern in which successive parasequences or sequences are stacked upwards and seawards, producing a net upward shallowing in marine and coastal settings.
Retrogradational stacking	A stacking pattern in which parasequences or sequences are stacked upwards and landwards, producing a net upward deepening in marine and coastal settings.

Table 1 (cont.)

Rheotrophic	A mire in which the groundwater is supplied primarily from subsurface flow.
Sequence boundary	Subaerial unconformity and correlative surfaces, generally thought to have chronostratigraphic significance, in which all rocks above the sequence boundary are younger than all rocks below the sequence boundary. Sequence boundaries are the upper and lower bounding surfaces of depositional sequences. Sequence boundaries form in response to a net loss of accommodation.
Simple coal	A coal recording one episode of mire formation and accumulation. Simple coals lack hiatuses and record one episode of wetting or drying of a mire.
Systems tract	Linkage of contemporaneous depositional systems, which are three-dimensional assemblages of lithofacies. Systems tracts are defined by their position within sequences and by their internal stacking pattern.
Taphonomy	Study of processes of fossil accumulation and preservation and how they affect information in the fossil record.
Time averaging	The incorporation of the biological remains of successive, noncontemporaneous populations of organisms into a single bed.
Transgressive systems tract (TST)	The second systems tract within a coastal depositional sequence, characterized by retrogradational stacking.
Well-drained paleosol	Ancient soil developed in association with a consistently or seasonally low water table, which facilitates the introduction of oxygen, increasing the rate of organic-matter oxidation and promoting the oxidation of iron and manganese. Typically identified by red, maroon, brown, or yellow (high chroma) color.

in elevation. Over shorter time scales as the alluvial plain responds to changes in regional base level (often a lake or a major axial river; i.e., Holbrook et al. 2006; Table 1), the basin will experience cyclic changes in elevation driven by changes in subsidence and uplift. Although changes in elevation may be small over short time scales in response to modest changes in base level, the total change in elevation over the history of a basin may be significant, up to a couple hundred meters, or more if the basin undergoes epeirogenic uplift or subsidence.

5 Stratigraphic Architecture of Nonmarine Strata

From the perspective of a stratigraphic paleobiologist, the critical issue is how the facies that preserve nonmarine depositional environments are distributed through the stratigraphic record. The central issue is whether facies occur randomly or if they are stratigraphically organized and predictable. If the latter is true, then stratigraphic patterns in the occurrence of species will reflect not only when species were extant and their environmental distribution (e.g., Lyson and Longrich 2011), but also the preservation of depositional environments, and the preservation of species within those environments (Loughney and Badgley 2017). As such, successfully interpreting patterns of fossil occurrence requires an understanding of depositional environments and how fossils are preserved in them (e.g., Wing and DiMichele 1995). When depositional environments are considered in relation to the spatial and temporal patterns of basin subsidence (and eustasy in coastal systems), the stratigraphic architecture of nonmarine systems is revealed — that is, the vertical and lateral distribution of sedimentary facies that contain the nonmarine fossil record.

5.1 Preservational Settings of Nonmarine Fossils

The preservation potential of major nonmarine fossil groups varies markedly among depositional environments (Fig. 7–9), a topic addressed comprehensively by Behrensmeyer and Hook (1992). Plants and animals may be widely distributed, but they do not have equal chances of being preserved in all environments in which they live at a given point in time. Although many nonmarine depositional environments are represented in the rock record, some are more common sites of fossil accumulation and preservation than others.

In depositional environments that preserve fossils, the composition and characteristics of those assemblages depend on the local physical and chemical conditions that govern the burial and preservation of organic remains (Figs. 9–10; Retallack 1984; Behrensmeyer and Hook 1992). Macrofloral remains (leaves, stems, roots, reproductive bodies, etc.) may be well preserved and abundant in lakes, abandoned channels, crevasse splays, poorly drained floodplains, mires (Figs. 7–8), and other environments where elevated water tables and reducing conditions inhibit microbial degradation (Fielding 1987; Gastaldo 1988; Spicer 1989; Gastaldo and Staub 1999; Gastaldo and Demko 2011; Wing 2005). These constraints are also critical for the formation and preservation of coal (McCabe 1984; McCabe and Parrish

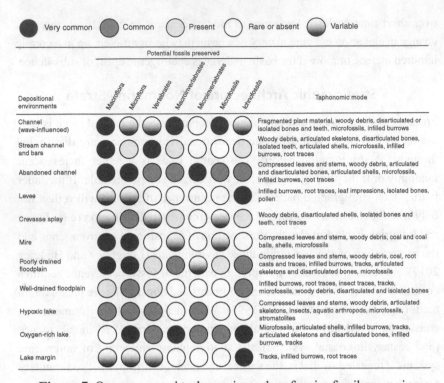

Figure 7. Occurrence and taphonomic modes of major fossil groups in nonmarine depositional environments, based on Behrensmeyer and Hook (1992).

1992; Bohacs and Suter 1997). Microflora (e.g., pollen, phytoliths) occur in a wide variety of settings, including those that preserve macroflora, but also in better-drained settings (Retallack 1984). Vertebrate remains may be common in channel lags, abandoned channels, and many lake settings, but they also occur in crevasse splays, floodplain deposits, and in coastal settings (Figs. 7–8). Because the preservation of bones and teeth is inhibited by acidic waters, they tend not to occur with plant fossils, as reducing pore waters are often acidic as well (Fig. 10; McCabe 1984; Retallack 1984; McCabe and Parrish 1992; DiMichele and Phillips 1994; Wilf et al. 1998; Gastaldo and Staub 1999). Alkaline pore waters also promote bone permineralization, such as in some well-drained paleosols (Bown and Kraus 1987; Behrensmeyer and Hook 1992; Badgley and Behrensmeyer 1995; Behrensmeyer et al. 1995), and in hydromorphic paleosols that were reducing but nonacidic (Loughney et al. 2011). Macroinvertebrate fossils are commonly preserved in coastal settings, in channel and abandoned-channel deposits, in deposits of oxygenated lakes (Figs. 7, 9), and places where alkaline pore waters permit the preservation of calcareous fossils (Fig. 10). Fossils of microinvertebrates (i.e., insects, spiders, aquatic arthropods, etc.) may have the narrowest range of

Figure 8. Examples of modern depositional settings in nonmarine systems. A:
Tidally influenced meandering river and floodplain, Whisky George Creek, near
Creels, Florida, USA; B: Channels and gravel bars of a braided river, Elbow River,
southwest of Braggs Creek, Alberta, Canada; C: Coalescing alluvial fans, near
Tonopah, Nevada, USA; D: Great Swamp, South Kingstown, Rhode Island, USA;
E: Point bar of a braided river, Elbow River, southwest of Braggs Creek, Alberta,
Canada; F: River scours in a degradational landscape, Mount St. Helens,
Washington, USA.

depositional settings in which they potentially occur and are predominantly rele-
gated to anoxic lake sediments, often restricted to deeper-water areas below the
oxygen-minimum zone (Martin 1999). Ichnofossils are common in many nonmarine and coastal settings and may occur in depositional environments that do not
preserve body fossils (Figs. 7, 9). Microfossils (ostracods, diatoms, etc.) also occur
in a variety of nonmarine deposits ranging from those of lakes to well-drained
floodplains (Fig. 8).

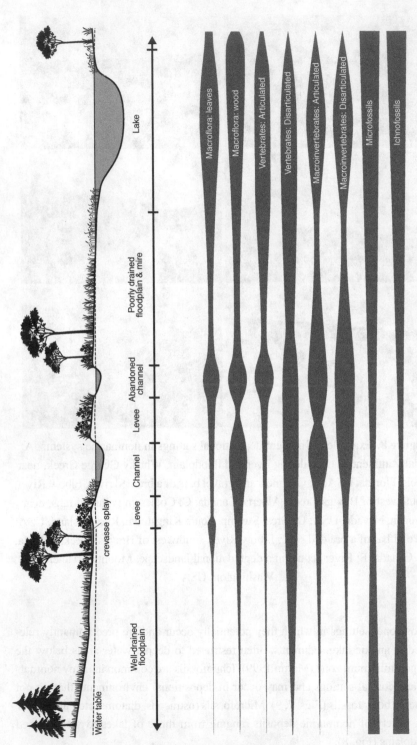

Figure 9. Schematic preservational potential of major fossil groups in fluvial and lacustrine settings in a seasonally wet setting. No vertical or lateral scale is implied. Modified from Loughney and Badgley (2020).

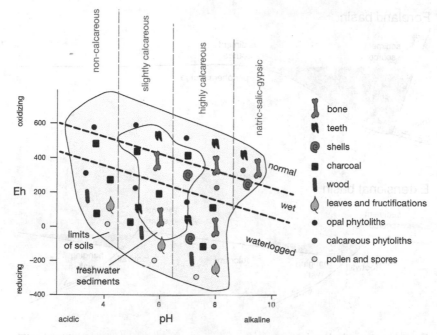

Figure 10. Geochemical (Eh and pH) controls on the preservation of major fossil groups. Modified from Retallack (1984).

The variable preservation of organisms or their body parts among depositional environments creates taphonomic gradients across landscapes (Fig. 9). These taphonomic gradients, in addition to biological gradients in species occurrence, control patterns of fossil occurrence, including taxonomic composition, the quality of preservation (e.g., diagnostic or nondiagnostic material), and the type of preservation (i.e., body fossils, ichnofossils, etc.; Fig. 9; Loughney and Badgley 2020).

Depositional environments with strong preservation potential are not arrayed randomly through the stratigraphic record. Understanding where they occur involves understanding the generation of stratigraphic architecture, from the formation of sedimentary basins to shorter-term controls on the rates of accommodation and sedimentation.

5.2 Spatial Variations in Basin Subsidence

Stratigraphic architecture in marine and nonmarine settings is controlled by the rates of accommodation and sedimentation (Catuneanu 2006). Sediment supply is controlled by numerous factors, including topography and bedrock type in the source area, as well as climate, which controls rates of weathering and river discharge. Autogenic processes of fluvial sedimentation exert

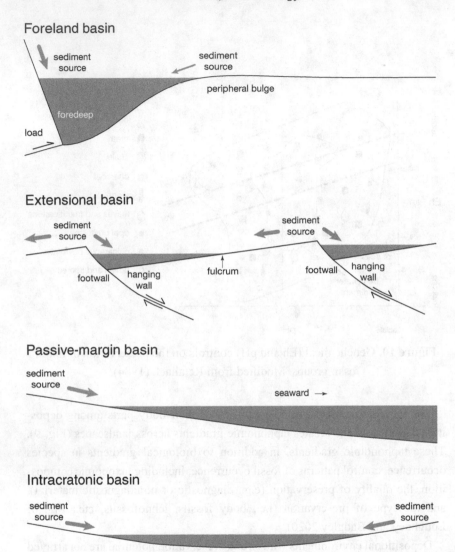

Figure 11. Subsidence and uplift within foreland, extensional (half-graben), passive-margin, and intracratonic basins. No consistent scale is implied among the basin types, owing to their greatly differing sizes. Areas experiencing subsidence are shown in gray.

a strong control on local sedimentation rates. Accommodation in both inland and coastal sedimentary basins is controlled tectonically by rates of uplift and subsidence. In coastal basins, eustatic sea level adds an additional component to accommodation. Patterns of subsidence vary among basin types, giving them characteristic stratigraphic architectures (Angevine et al. 1990; Allen and Allen 2005).

In foreland basins (Fig. 1), subsidence is driven primarily by crustal flexure caused by the weight of the adjoining fold-and-thrust-belt, termed the load (Fig. 11). Subsidence rates increase towards the load, creating a structural depression known as a foredeep (DeCelles and Giles 1996). The foredeep is flanked distally by a broad and gentle uplift called the peripheral bulge (Pope et al. 2009). Foreland basins are typically tens to several hundreds of kilometers wide, hundreds of kilometers long, and may persist for tens to hundreds of millions of years (Jordan 1995). Most sediment introduced into a foreland basin is produced by erosion of the fold-and-thrust belt, but lesser amounts of sediment can also be supplied from the peripheral bulge and the continental interior.

In extensional basins (Fig. 2), subsidence is driven by stretching of the lithosphere (Fig. 11). Extensional basins are bounded by normal faults on one (asymmetric, half-graben) or both (symmetric, full-graben) margins. Fault geometries may be complicated, but half-grabens are more common (Rosendahl 1987). In large extensional regions, such as the Basin and Range province of western North America, imbricated fault blocks are tilted along detachment faults, which form a series of half-grabens bounded by normal faults (Withjack et al. 2003). Although individual half-grabens may be on the order of 10 km wide, they lie within broader extensional systems that can be hundreds of kilometers wide and thousands of kilometers long. Along each normal fault, subsidence occurs on the hanging wall, and uplift of the footwall creates a fault scarp (Fig. 11; Leeder 1995). Subsidence and uplift rates are greatest on opposite sides of the footwall scarp and decrease away from it (Sharp et al. 2000). Tilting of a fault block between two normal faults creates a point of zero subsidence, called the fulcrum (Fig. 11; Leeder 1995). Subsidence and sedimentation increase towards the basin on the hanging wall side, with increasing uplift and erosion updip on the footwall side. Sediment is shed into extensional basins through transverse drainages emanating from the footwall or the hanging wall and in axial drainages that develop parallel to the fault scarps. Hanging-wall catchments (drainage basins) are often larger than those on the footwall and provide a greater supply of sediment (Leeder and Gawthorpe 1987; Gawthorpe and Leeder 2000). The deposits of axial streams are thickest near the fault scarp where subsidence rates are highest.

In passive-margin (Fig. 3) and intracratonic basins, subsidence is driven by cooling of the lithosphere following an initial stretching event, and rates of subsidence decline exponentially over time (Pitman 1978; Armitage and Allen 2010). In passive-margin basins, subsidence rates increase approximately linearly away from the continent, and in intracratonic basins, subsidence rates increase towards the center of the basin (Fig. 11).

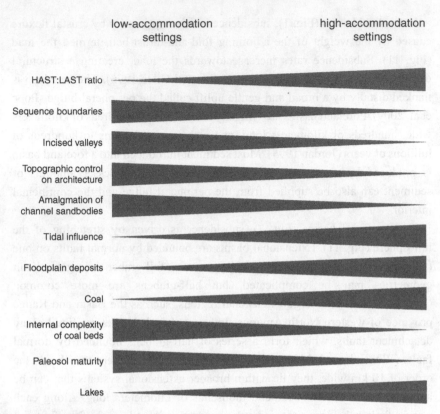

low-accommodation
settings   high-accommodation
                                                  settings

HAST:LAST ratio

Sequence boundaries

Incised valleys

Topographic control
on architecture

Amalgmation of
channel sandbodies

Tidal influence

Floodplain deposits

Coal

Internal complexity
of coal beds

Paleosol maturity

Lakes

Figure 12. Expected stratigraphic features in settings ranging from low accommodation to high accommodation. Based in part on Catuneanu (2006).

5.3 Accommodation Settings

The distinction between regions of high versus low subsidence in nonmarine basins is a first-order control on the stratigraphic architecture of nonmarine strata (Catuneanu 2006), although other factors such as autogenic dynamics are also important (e.g., Bryant et al. 1995; Mackey and Bridge 1995; Hajek and Heller 2012; Chamberlin and Hajek 2015). Nonmarine stratigraphers have simplified the spectrum of accommodation rates by emphasizing two end-member cases: a high-accommodation setting and a low-accommodation setting (Arnott et al. 2002; Zaitlin et al. 2002; Catuneanu 2006).

High-accommodation settings are characterized by relatively thick deposits, directly reflecting the high subsidence rate (Fig. 12; Table 2). They are dominated by thick and extensive fine-grained floodplain deposits, with single-story channel sand bodies that are vertically and laterally isolated from one another and separated by floodplain deposits. Paleosols in these thick floodplain deposits are immature (Table 1) because they are buried rapidly. In areas

Figure 13. Outcrops of nonmarine facies that host fossil assemblages. A: Conglomerates deposited by alluvial fans, Cretaceous, Utah, USA; B: Multistory channel sandstones, Crab Orchard Mountains Group, Pennsylvanian, Tennessee, USA; C: Coal beds in the Breathitt Group, Pennsylvanian, Kentucky, USA; D: Abandoned-channel deposit in the Horseshoe Canyon Formation, Cretaceous, Alberta, Canada; E: Alternating high-accommodation (HAST) and low-accommodation (LAST) systems tracts in the Castlegate Sandstone, Cretaceous, Utah, USA; F: Fine-grained sandy deposits of a shallow lake system, Green River Formation, Eocene, Wyoming, USA; G: Hydromorphic paleosol in the Chinle Formation, Triassic, Arizona, USA; H: Mature, well-drained paleosol in the Chinle Formation, Triassic, Arizona, USA.

Table 2 Accommodation settings in nonmarine systems, adapted from Catuneanu (2006).[a]

Feature	Low-accommodation Settings	High-accommodation Settings
A:S ratio	Low	High
Thickness	Thin	Thick
Systems tracts	Dominated by LAST in inland settings, HST commonly missing in coastal settings	Dominated by HAST in inland settings, all systems tracts in coastal settings
Sequence-bounding unconformities	Multiple, closely spaced	Rare
Incised valleys	Multiple and common, often with substantial truncation of underlying strata	Rare, often with little truncation of underlying strata
Underlying topography	Enhanced control on architecture	Weak control on architecture
Channel sandstones	Generally amalgamated, may be less so near maximum flooding surface or within HAST	Commonly single-story to weakly amalgamated, more amalgamated above and below sequence-bounding unconformities
Tidal influence	Less common, but more likely in coastal settings near maximum flooding surface	Common, especially in coastal settings near maximum flooding surface
Floodplain deposits	Rare, more likely developed within TST in coastal settings and HAST in inland settings	Abundant, most dominant in TST in coastal settings and HAST in inland settings
Coals	Typically absent, but with multiple internal hiatuses where present	Abundant, commonly thick, with fewer internal hiatuses
Paleosols	Uncommon, but those preserved tend to be well-drained and mature.	Thin, widely spaced, and immature. Tend to be hydromorphic and organic-rich.

[a] LAST: low-accommodation systems tract; HAST: high-accommodation systems tract; HST: highstand systems tract; TST: transgressive systems tract.

where the water table is high or intersects the land surface, hydromorphic paleosols may develop (Fig. 13, G; Table 1). In high-accommodation settings where the climate is wet, sediment input is limited, and the other physicochemical requirements for peat are met (e.g., Gastaldo 2010), such as in floodplain areas far from fluvial channels, coals are common, thick, and simple in their internal anatomy (Table 1). Large lakes or numerous small lakes may form in high-accommodation settings. Sequence-bounding unconformities tend to be rare and widely spaced (Fig. 12; Table 2), because high subsidence rates make it difficult to achieve the negative rates of accommodation necessary to form an unconformity.

Low-accommodation settings have thin stratigraphic records with relatively few or no floodplain deposits. They are dominated by highly amalgamated multistory or multilateral channel sand bodies, in which channels truncate older channel deposits. This truncation can be caused by lateral migration of channels and by avulsion of channel belts (Hajek and Wolinsky 2012). Where floodplain deposits are present, they typically contain paleosols that are mature (Table 1) and oxidized, reflecting a relatively low water table (Figs. 12, 13, H; Table 2). Coal is usually absent from low-accommodation settings, but where present, coal beds are compound with multiple internal discontinuities that reflect a complicated depositional history. Sequence-bounding unconformities are common and closely spaced in low-accommodation settings, and they may incise deeply into older strata (Fig. 12; Table 2).

Although settings within a basin are often described as either low-accommodation or high-accommodation, accommodation varies over a spectrum of conditions (Fig. 12). As a result, depositional settings might be described as intermediate (e.g., Wadsworth et al. 2002), hyper-high, or hyper-low accommodation (Kidwell 1993 presents a similar argument for marine systems). Regardless of the terms that are applied, the implication is the same: by controlling the preservation of various depositional environments, the rate of accommodation controls the nonmarine stratigraphic record and its fossil record.

5.4 Temporal Variations in Basin Subsidence

Most sedimentary basins experience a decline in subsidence rates through their history (Fig. 14). For example, foreland basins have high initial rates of subsidence as thrust loads are emplaced. These are followed by declining rates of subsidence and eventually by uplift as thrusting ceases and erosion of the thrust loads outpaces their emplacement (Heller et al. 1988). Although foreland basins may undergo multiple cycles of loading and quiescence, they display a net long-term decline in subsidence rate as thrust emplacement terminates and the thrust loads are eroded. This long-term decline typically takes place over the lifespan of the basin, which is typically tens of millions of years (Angevine et al. 1990; Allen and Allen 2005). Similarly, rift basins have high initial rates of subsidence during extension, followed by a decrease in subsidence rates as stretching ceases (Fig. 14; Bridge 2003). Like foreland basins, rift systems typically last for a few tens of millions of years (Leeder 1995). Passive-margin and intracratonic basins also have long-term declines in subsidence rate, owing to the exponential law of cooling-generated subsidence. In contrast to foreland and extensional basins, passive-margin and intracratonic basins are longer lived, lasting hundreds of millions of years.

Declining subsidence rates cause nonmarine basins to evolve towards lower-accommodation conditions over time (Fig. 14). With their rapid initial subsidence rates, foreland and extensional basins may start as high-accommodation settings, but they evolve toward low-accommodation settings. In contrast, the nonmarine portions of passive-margin and intracratonic basins start as low-accommodation settings and evolve towards hyper-low or even negative accommodation, in which unconformities cannibalize much of the record that was previously deposited.

Declining subsidence rates in basins may also be accompanied by increased sediment supply as uplands erode, additionally causing fluvial systems to prograde progressively into the basin over time (Fig. 15). This progradation causes a net stratigraphic trend of increasing elevation at any location, which will be reflected in the composition of fossil communities.

Basin types display two distinct spatial relationships between subsidence rates and the location of sediment sources (Fig. 15). In foreland and extensional basins, sediment is introduced to the basin by rivers where subsidence rates are greatest, which allows thick accumulations of nonmarine sediment. In passive-margin and cratonic basins, sediment enters where subsidence rates are slowest, causing accumulations of nonmarine strata to be thinner. In addition, subsidence rates of extensional and foreland basins are greater than passive-margin and cratonic basins (Fig. 11; Angevine et al. 1990; Allen and Allen 2005). Accordingly, nonmarine deposits in passive-margin and intracratonic basins are typically thinner and less extensive than those in extensional and foreland basins.

As such, a greater proportion of the nonmarine fossil record is found in foreland and extensional basins. However, foreland and extensional basins are smaller than passive-margin and intracratonic basins, and they are shorter-lived by an order of magnitude (Holland 2016). As a result, the basins with the majority of the nonmarine fossil record tend to represent relatively small geographic areas and short spans of geologic time.

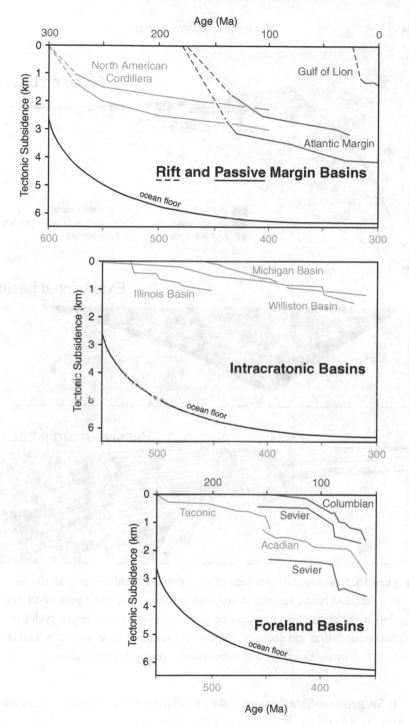

Figure 14. Subsidence histories of representative extensional (rift; dashed line) and passive-margin basins (solid line), intracratonic basins, and foreland basins. All plots are to the same scale, with subsidence of oceanic crust shown for scale. Blue curves follow the time scale at the bottom of each figure, and red curves follow the time scale at the top. Adapted from Angevine et al. (1990).

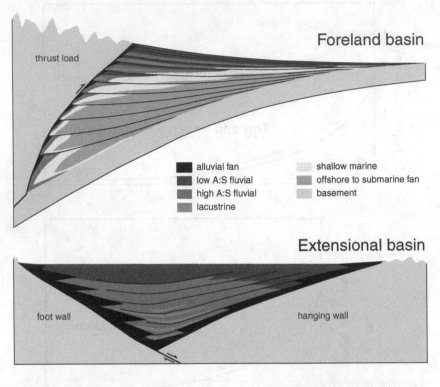

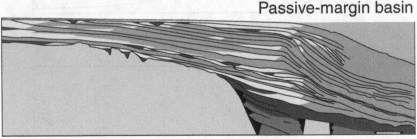

Figure 15. Schematic illustrations of stratigraphic architecture over the history of a foreland basin, an extensional half-graben basin, and a passive-margin basin. Actual architecture may be more complicated, owing to cycles of accommodation and sedimentation over a range of time scales, as well as synsedimentary and subsequent structural deformation.

6 Sequence-Stratigraphic Architecture of Nonmarine Basins

Sequence stratigraphy is the study of how changes in the rates of accommodation (i.e., tectonic subsidence and eustasy) and sedimentation control the architecture of the stratigraphic record. In particular, sequence-stratigraphic models have emphasized how cycles in the rate of accommodation produce unconformity-bounded packages of strata called depositional sequences. Owing to predictable changes in the relative rates of accommodation and sedimentation, depositional sequences are divided into packages of strata that have a characteristic stratigraphic position and internal anatomy, called systems tracts (Van Wagoner et al. 1988; Catuneanu 2006; see Table 1 for definitions of common sequence-stratigraphic terms). For non-marine strata, separate frameworks have been proposed for coastal settings influenced by sea level (Fig. 16, left) and inland settings that operate independently of sea level (Fig. 16, right). Systems tracts for distributive fluvial systems may be a third framework but have yet to be fully developed.

6.1 Coastal Systems Tracts

The classic marine systems tracts (Van Wagoner et al. 1988; Hunt and Tucker 1992; Catuneanu 2006) can be extended landward to where marine and non-marine deposits interfinger (Figs. 17–18; Wright and Marriott 1993; Shanley and McCabe 1995; Bohacs and Suter 1997). In these settings, it is generally

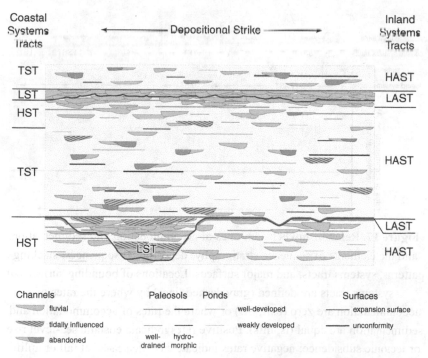

Figure 16. Stratigraphic architecture of fluvial settings, based on Wright and Marriott (1993) and Atchley et al. (2004), with coastal systems-tract model (left; Van Wagoner et al. 1990) and inland systems-tract model (right; Martinsen et al. 1999). Cross-section is along depositional strike and does not imply any particular basin type.

Elements of Paleontology

thought that eustasy is the primary source of variations in accommodation and, therefore, the agent producing depositional sequences. The effects of eustatic change are greatest at the shoreline and decrease landward, and the extent of the backwater region affected by sea level is a function of channel slope and tidal range. Backwater effects can reach kilometers to tens of kilometers from the shoreline on rivers with steep profiles (e.g., Eel and Colorado rivers) and for hundreds to thousands of kilometers on rivers with gentle gradients (e.g., Mississippi and Amazon rivers; Blum et al. 2013). These distances are upstream of the landward limits of tidal influence or saltwater intrusion, both of which can extend far inland from the river mouth.

A depositional sequence begins with the sequence boundary (Fig. 17–18). In depositionally updip areas, this is manifested as a subaerial unconformity, which forms by the incision of rivers and the development of mature paleosols on interfluves (Fig. 8, F). The erosional surface is commonly expressed as a high-relief surface with the formation of deep incised valleys, but it can also be

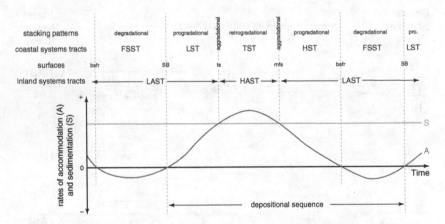

Figure 17. Relationship between rates of accommodation and sedimentation and the elements of sequence stratigraphy: depositional sequences, stacking patterns, systems tracts, and major surfaces. Locations of bounding surfaces of systems tracts are defined (gray dashed lines) by where the rates of accommodation are zero (bsfr, SB) or where the rates of accommodation and sedimentation are equal (ts, mfs). Positive rates indicate eustatic sea-level rise or tectonic subsidence; negative rates indicate eustatic sea-level fall or uplift. SB: sequence boundary; ts: transgressive surface; mfs: maximum flooding surface; bsfr: basal surface of forced regression; LST: lowstand systems tract; TST: transgressive systems tract; HST: highstand systems tract; FSST: falling-stage systems tract; LAST: low-accommodation systems tract; HAST: high-accommodation systems tract.

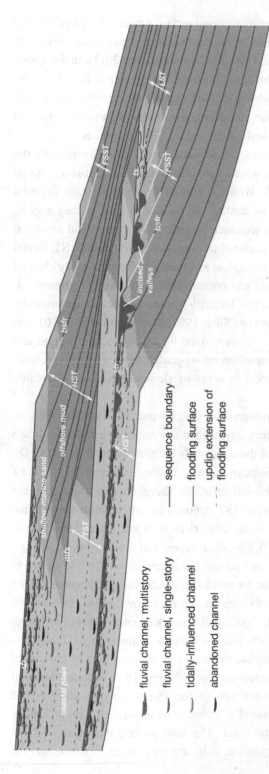

Figure 18. Schematic cross-section along depositional dip, showing distribution and character of nonmarine strata within a depositional sequence developed in a coastal basin. Note relative abundance of single-story vs. multistory channels, fluvial channels, tidally influenced channels, and abandoned channels among systems tracts and with distance from the coast.

a widespread and relatively low-relief erosional surface. On the interfluves between river valleys, protracted weathering of the land surface creates mature and well-drained paleosols (Fig. 16; Kraus 1999; McCarthy and Plint 2013). In the framework of Gastaldo and Demko (2011), in which the land surface of fluvial systems alternates between states of degradation (erosion), stasis (pedogenesis), and aggradation (accumulation), the formation of a sequence boundary occurs during a period dominated by degradation in incised valleys and stasis on interfluves.

In depositionally downdip areas, the sequence boundary is overlain by the lowstand systems tract (LST), which consists of stacked multistory fluvial channel sand bodies (Fig. 13, B) with limited or no floodplain deposits (Fig. 18). These deposits may be confined to incised valleys, or they may be regionally extensive where the sequence boundary is a widespread erosional surface. As the rate of eustatic sea level rise quickens through the LST, fluvial systems increasingly preserve floodplain sediments with single-story channel bodies. Thick, widespread coals and organic-rich muds may be present initially, but they become increasingly laterally limited as rates of accommodation increase (Ryer 1983; Bohacs and Suter 1997; Wadsworth et al. 2010). The land surface during the LST is characterized by a shift to aggradation and stasis, with an increasing proportion of aggradation to stasis over time, although degradation may persist in areas of slower subsidence (Gastaldo and Demko 2011).

The LST is overlain by the transgressive systems tract (TST), which consists in nonmarine areas of single-story fluvial channels encased in thick floodplain deposits (Figs. 16, 18). Many of these channels are abandoned (Fig. 13, C, D), owing to frequent avulsion. Stratigraphically upward, channels in the TST typically show increasing tidal influence, as shoreline transgression takes place. In coastal areas, nonmarine TST deposits are often capped by marine deposits. Because the coastal water table rises with sea level (Shanley and McCabe 1994; Blum et al. 2013), floodplain deposits of the TST have immature hydromorphic paleosols and small ponds. Coals tend to be thick in the early TST as rates of peat accumulation are able to keep pace, but they become thinner as peat accumulation is unable to keep up and mires are flooded (Bohacs and Suter 1997; Wadsworth et al. 2002; Holz and Kalkreuth 2004). Although siliciclastic marine TST deposits tend to be thin, nonmarine TST deposits tend to be thick, as sediment supplied to the basin is captured first by the coastal plain, particularly in areas close to river channels (Fig. 18; Wright and Marriott 1993). Within any vertical column through the TST, nonmarine strata record progressively more coast-proximal settings, corresponding to progressively lower elevations on the coastal plain. The land surface during the TST is dominated by periods of aggradation, with shorter periods of stasis, and the dominance of aggradation reaches a peak near the middle of the TST (Gastaldo and Demko 2011).

The TST is overlain by the highstand systems tract (HST). In areas of nonmarine strata, the HST displays an upward transition to more multistory channel sand bodies and fewer single-story, abandoned, and tidally influenced channel sandstones (Figs. 16, 18). This transition is accompanied by a gradual decrease in the thickness of floodplain deposits. Progradation of this system leads to a lower water table, forming better-drained and more mature paleosols and fewer ponded areas. Provided that the climatic and substrate conditions are conducive for peat formation, coals tend to be thick and widespread in the earliest HST (Bohacs and Suter 1997; Wadsworth et al. 2002; Holz and Kalkreuth 2004). They also tend to be thinner and more isolated upward within the HST as decreasing rates of accommodation limit the amount of peat accumulation. Nonmarine HST deposits tend to be thinner than those of the TST owing to declining rates of accommodation (Fig. 18). Within any vertical column through the HST, nonmarine strata record positions that are progressively farther from the coast and, therefore, at higher elevations. The land surface during the HST undergoes more frequent and longer periods of stasis relative to aggradation (Gastaldo and Demko 2011).

The HST is succeeded by the falling-stage systems tract (FSST). Because of the negative rates of accommodation (Fig. 17), any deposits tend to be thin and are the first to be eroded during the formation of the growing subaerial unconformity (Fig. 18). For this reason, early sequence-stratigraphic frameworks included FSST deposits as part of the late HST (Van Wagoner et al. 1988, 1990). Because the fall in sea level creates a scarp at the former shoreface, the coastal gradient is steepened, causing incision and valley formation rather than deposition in formerly coastal nonmarine areas. Consequently, the few nonmarine deposits of the FSST are usually limited to isolated terrace deposits on the walls of incised valleys (Blum et al. 2013). They are also deeply weathered and typically capped by mature, well-drained paleosols. During the FSST, the land surface is dominated by degradation of channel courses and stasis on interfluves (Gastaldo and Demko 2011).

This architecture changes predictably along depositional dip (Fig. 18). In depositionally downdip settings, nonmarine deposits may be limited to the LST. These deposits are bounded sharply below by the sequence boundary and overlie marine deposits, and they are bounded sharply above by the transgressive surface and overlain by marine strata. In progressively more updip settings, nonmarine strata extend higher into the TST, and these may ultimately be overlain by marine strata. In the HST, these marine strata pass conformably upwards into nonmarine strata. In other words, a wedge of marine strata separates the nonmarine strata in the underlying TST from those in the overlying HST. This marine wedge thins and eventually disappears depositionally updip, beyond which the entire TST and HST consist of nonmarine strata (Fig. 18). The upper and lower boundaries of this marine wedge differ, with an abrupt lower boundary lying at a marine flooding surface, and a conformable upper boundary marking the diachronous passage from shallow-marine strata into coastal nonmarine strata. There is a strong potential for each of these time-transgressive contacts to be correlated erroneously along depositional dip in non-sequence-stratigraphic studies (see Figures 17 and 18 of Van Wagoner et al. 1990).

6.2 Inland Systems Tracts

In coastal areas so far inland that they are beyond the influence of eustasy (Swenson 2005) and in inland basins isolated from the ocean, it is impossible to use the classic marine systems tracts (i.e., LST, TST, HST, FSST). Instead, two systems tracts are recognized, the high-accommodation and low-accommodation systems tracts (Figs. 13, E, 16, right, 19; Table 3; Martinsen et al. 1999; Boyd et al. 2000; Catuneanu 2006). These systems tracts are closely analogous to the high-accommodation and low-accommodation settings.

Low-accommodation systems tracts (LAST) form when the rate of accommodation is low relative to the rate of sedimentation. Under these conditions, the floodplain aggrades slowly, channel avulsion is infrequent, and lateral-accretion deposits characterize the system. As a result, LAST deposits are thin and dominated by multistory and multilateral channel sand bodies (Fig. 16; Table 3). Fine-grained floodplain deposits are, therefore, a minor component of the LAST and may be absent altogether (Wright and Marriott 1993). Where floodplain deposits are preserved, they often contain mature well-drained paleosols, reflecting longer exposure and weathering (Kraus 1999). The tops of channel sand bodies may also be pedogenically modified. Low water tables inhibit the formation of rheotrophic peats (Table 1), and low rates of accommodation inhibit the preservation of both rheotrophic and ombrotrophic peats (Table 1). As a result, coal deposits are typically absent in the LAST (Bridge 2003; Catuneanu 2006).

The high-accommodation systems tract (HAST) forms when rates of accommodation are high. Rapid floodplain aggradation promotes frequent channel avulsion (Bryant et al. 1995), and the resulting HAST deposits are typically dominated by thick successions of fine-grained floodplain deposits with isolated, single-story channel sandstones (Fig. 16; Table 3; Catuneanu 2006). Abandoned-channel deposits have a high potential for preservation and can be

Table 3 Systems tracts of inland nonmarine deposits, adapted from Martinsen et al. (1999) and Catuneanu (2006).

Feature	Low-accommodation Systems Tract (LAST)	High-accommodation Systems Tract (HAST)
A:S ratio	Low	High
Thickness	Generally thin, but can be thick if low-accommodation conditions are persistent	Often thick, but can be thin if truncated by an unconformity, or if high-accommodation conditions were short-lived
Floodplain deposits	Rare and thin	Common, thick
Coal	Rare, thin, usually with multiple hiatuses and wetting and drying trends	Common, thick, typically lacking hiatuses and with single wetting or drying trend
Paleosols	Common, typically mature and well-drained	Few, typically thin, immature, may be hydromorphic
Channel sandstones	Amalgamated, multistory, and multilateral	Isolated, single-story, may be tidally influenced
Distribution	Variable; may be laterally restricted and confined to an incised valley, or may form a regional sheet-like geometry	Typically widespread and of regional extent
Upper bounding surface	Upper bounding surface commonly a sharp expansion surface, but it may also be a sequence-bounding unconformity. In some cases, the sequence-bounding unconformity may lie within the LAST.	Upper bounding surface often gradational to overlying HAST; may also lie at a sequence-bounding unconformity.
Lower bounding surface	Lower bounding surface commonly a sequence-bounding unconformity, but in some cases is gradational with underlying HAST.	Lower bounding surface is commonly a sharp expansion surface with underlying LAST. Base may also coincide with a sequence-bounding unconformity.

common in the HAST (Fig. 16). Steady sediment deposition on the floodplain impedes pedogenesis, and paleosols tend to be immature (Wright and Marriott 1993; Kraus 1999). High water tables promote the formation of hydromorphic paleosols and ponds. High water tables also promote the development of rheotrophic peats if the rate of organic matter input is high. The resulting coals may be common, thick, and simple. High aggradation rates allow for the preservation of rheotrophic as well as ombrotrophic coals. If the climate is humid and sediment supply is limited, whether by weathering (Cecil and Dulong 2003) or upstream trapping of sediment, large and potentially deep lakes may form (Fig. 13, F). In these cases, much of the HAST may be thick lacustrine deposits.

As is true for the high- and low-accommodation settings, these systems tracts are artificial divisions of a continuum of possible architectural patterns that may develop over the history of a basin (Fig. 19). As such, intermediate as well as more exaggerated cases (i.e., hyper-HAST or hyper-LAST) may form, depending on the rate of accommodation (Boyd et al. 2000).

A depositional sequence in an inland area begins with a sequence boundary that marks a period of negative accommodation and landscape degradation (Fig. 16; Gastaldo and Demko 2011). As rates of accommodation slowly increase, stasis and aggradation characterize the landscape, with a progressive increase in aggradation over stasis. As a result, a LAST forms above the sequence boundary, and it may initially be restricted to incised valleys (Fig. 16; Wright and Marriott 1993). As rates of accommodation continue to increase, and aggradation dominates over stasis, a HAST develops. If this increase in the rate of accommodation is gradual, the contact between the underlying LAST and the overlying HAST is gradational. The contact will be characterized by a progressive upward increase in floodplain deposits and a decrease in channel sand-body connectedness. These changes lead to an increasing occurrence of isolated single-story channel bodies over multistory or multilateral channel bodies. If the increase in accommodation is rapid, the contact of the HAST on the LAST is sharp (e.g., Rogers et al. 2016) and is called an expansion surface (Martinsen et al. 1999; Table 1). As rates of accommodation progressively decrease through the HAST, and stasis dominates over aggradation, the HAST passes gradationally upwards into a LAST, capped by a sequence boundary and the LAST of the following sequence. If the sequence boundary is marked by widespread erosion, the LAST underlying the sequence boundary may be erosionally removed, causing the sequence boundary to be a sharp contact where the LAST overlies a HAST (e.g., Martinsen et al. 1999).

Regardless of the placement of the sequence boundary, alternations of LAST and HAST (or similar architectures with different names) are common in non-marine sediments (e.g., Currie 1997; Martinsen et al. 1999; Gani et al. 2015;

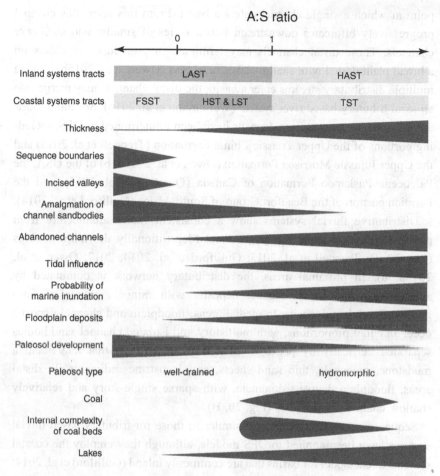

Figure 19. Relationships of the coastal and inland systems-tracts models to the accommodation to sedimentation ratio (A:S), with their corresponding expressions in the nonmarine stratigraphic record. Based in part on Catuneanu (2006).

Scherer et al. 2015; Wanas et al. 2015; Campo et al. 2016; Loughney and Badgley 2017). In low-accommodation settings, depositional sequences may consist entirely of a LAST bounded above and below by sequence boundaries.

6.3 Distributive Fluvial Systems

Although tributary fluvial systems (such as the Mississippi River and its tributary system) are well-known and are the basis of many depositional models, distributive fluvial systems (DFS), in which rivers bifurcate downstream, have been increasingly described in foreland and extensional basins (Hartley et al. 2010; Weissmann et al. 2010, 2015). A DFS radiates from the point at which a single channel enters a basin. From this apex, this channel progressively bifurcates downstream into a series of smaller and shallower channels. These distal channels may terminate in sand lobes or sheets on a broad plain dotted with shallow lakes or playas (Owen et al. 2015b). Where multiple distributary streams enter a basin, the distal channels may merge into an axial tributary-based river system (Weissmann et al. 2010). Several important fossil-bearing nonmarine deposits have been reinterpreted as DFS, including portions of the Upper Triassic Chinle Formation (Trendell et al. 2013) and the Upper Jurassic Morrison Formation (Owen et al. 2015a, b) of the USA, the Paleocene Paskapoo Formation of Canada (Quartero et al. 2015), and the Permian portion of the Beaufort Group of South Africa (Gulliford et al. 2014).

Distributive fluvial systems show a consistent facies architecture from proximal (depositionally updip) to distal (depositionally downdip) settings (Fig. 20, B; Trendell et al. 2013; Gulliford et al. 2014, 2017; Owen et al. 2015a, b). In proximal areas, the distributary network is dominated by amalgamated multistory channel deposits with minor amounts of fine-grained floodplain deposits. In medial areas, floodplain and channel deposits occur in equal proportions, with multistory and isolated channel sand bodies separated vertically by packages of floodplain deposits that may include mudstone, paleosols, thin sand sheets, and lacustrine sediments. In distal areas, floodplain deposits dominate, with sparse single-story and relatively shallow channel sand bodies (Fig. 20, B).

Sequence-stratigraphic concepts similar to those for tributary-based fluvial systems have been applied to DFS models, although they employ the coastal systems tracts model for basins that are commonly inland (Gulliford et al. 2014; Owen et al. 2015a, b). For example, the LST overlies the sequence-bounding unconformity and is characterized by multistory sandstone bodies. These deposits pass upwards into progressively more floodplain-dominated deposits and single-story channel bodies in the TST, with more frequent indicators of poor drainage and elevated water tables near the maximum flooding surface equivalent. The HST is characterized by an upward increase in the channel- to floodplain-deposit ratio, as well as an upwards increase in the proportion of multistory or multilateral channel sandstones. In large part, these vertical trends are interpreted to result from the lateral translation of the proximal, medial, and distal belts of the DFS in response to changes in accommodation (Gulliford et al. 2014). In other words, a high rate of accommodation causes a retreat of the proximal and medial belts towards the apex of the DFS, accompanied by an expansion of the distal belt across the basin. Low rates of accommodation are accompanied by progradation of the proximal and medial belts, displacing the distal regions farther down depositional dip (Gordon and Heller 1993).

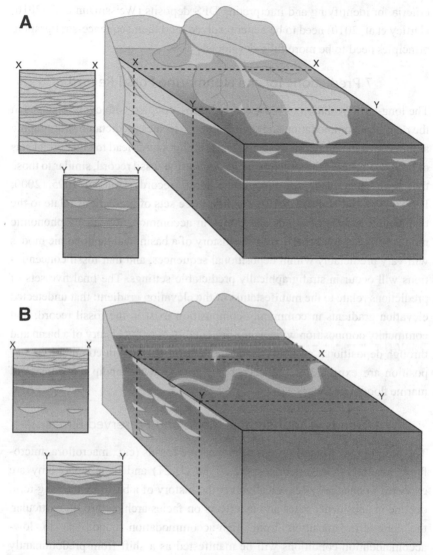

Figure 20. Schematic cross sections of fluvial architecture in distributary fluvial systems (A) and tributary fluvial systems (B). Light green = high accommodation; dark green = low accommodation.

Temporal changes in accommodation can produce similar changes in stacking patterns without regional displacements in the locations of the proximal, medial, and distal facies belts (Owen et al. 2015a).

Although DFS certainly occur in modern fluvial drainages (Weissmann et al. 2010, 2015), their ubiquity may be overstated, and their dominance in the rock record is debated (Sambrook Smith et al. 2010; Fielding et al. 2012). The criteria for identifying and interpreting DFS deposits (Weissmann et al. 2010; Hartley et al. 2010) need to be better resolved, and their sequence-stratigraphic principles need to be more fully developed.

7 Predictions for the Nonmarine Fossil Record

The long-term evolution of sedimentary basins, coupled with cyclic changes in the rates of accommodation and sedimentation, creates predictable stratigraphic architectures in nonmarine systems. These architectures lead to eight broad sets of predictions about the structure of the nonmarine fossil record, similar to those that have been made for the marine fossil record (Holland 1995, 2000; Patzkowsky and Holland 2012). The first three sets of predictions relate to the taphonomic effects of trends and cycles in accommodation: that taphonomic modes will vary predictably over the history of a basin, that taphonomic modes will vary predictably within depositional sequences, and that fossil concentrations will occur in stratigraphically predictable settings. The final five sets of predictions relate to the manifestation of the elevation gradient: that undetected elevation gradients in community composition exist in the fossil record, that community composition will change predictably over the history of a basin and through depositional sequences, and that abrupt changes in community composition are expected at sequence boundaries and the updip equivalents of marine flooding surfaces.

7.1 Trends in Taphonomic Mode and Preserved Biotas

The occurrence of major groups of nonmarine fossils (e.g., macroflora, microflora, invertebrates, vertebrates, etc.; Figs. 7, 21) and their taphonomy are expected to change systematically over the history of a basin by the long-term decline in subsidence rates and its effects on facies architecture. In particular, the long-term transition from high-accommodation conditions to low-accommodation conditions will be manifested as a shift from predominantly floodplain, lake, and single-story channel deposits to an increasing dominance of multistory or multilateral channel bodies. This transition will also include a decline in the amount of preserved floodplain and lake deposits. This long-term trend over the history of a sedimentary basin has three implications for the nature of the nonmarine fossil record.

First, the preservation quality of nonmarine fossil assemblages is expected to decline over the history of a basin. In the early high-accommodation phase of a basin, facies architecture will be dominated by floodplain, lake, abandoned-channel, and mire deposits (Fig. 21). These facies tend to have high preservation potential and tend to preserve autochthonous to parautochthonous assemblages.

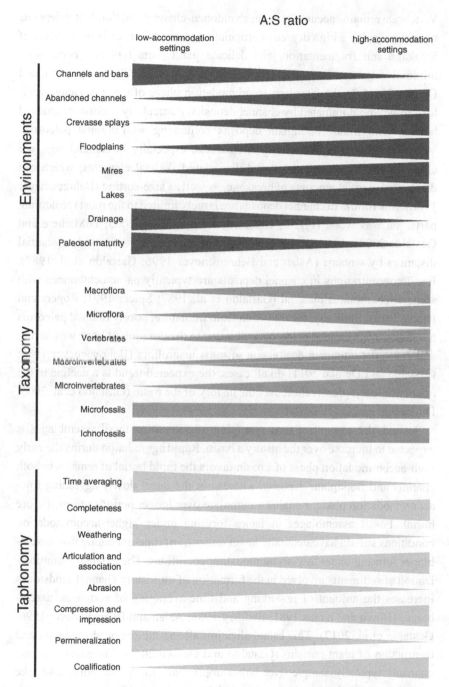

Figure 21. Expected variations of the dominant depositional environments, major taxonomic groups, and taphonomic patterns as a function of the accommodation to sedimentation ratio (A:S). All distributions are generalized and meant to show only first-order patterns of increase, decrease, or no trend.

Vertebrate remains accumulating in abandoned-channel and floodplain deposits typically display a high degree of articulation, association, and low amounts of breakage and fragmentation, and delicate plant parts typically occur with minimal fragmentation (Fig. 7; Behrensmeyer et al. 2000; Gastaldo and Demko 2011). In the later low-accommodation phase of a basin, facies architecture will be dominated by channel deposits, particularly multistory channel bodies, with minor floodplain deposits containing well-drained paleosols. Vertebrate assemblages forming in channels, especially in channel lags, are characterized by disarticulated and dissociated skeletal elements, which can display substantial amounts of breakage, as well as size-sorting (Behrensmeyer 1988). Macroflora in channel deposits are largely limited to the most recalcitrant parts, such as wood (Fig. 7; Greenwood 1991; Wing 2005; DiMichele and Gastaldo 2008). Although bones and macroflora can be transported substantial distances by streams (Aslan and Behrensmeyer 1996; Gastaldo et al. 1987), fossil concentrations in channel deposits are typically parautochthonous with some allochthonous material (Gastaldo et al. 1987; Spicer 1991; Rogers and Brady 2010). Similarly, the transition from immature, poorly drained paleosols to mature, well-drained paleosols will increase the duration of bone weathering and cause the complete destruction of most macroflora (Behrensmeyer 1978; Gastaldo and Demko 2011). In all cases, the expected trend is a decline in the quality of fossil preservation over the history of the basin (Gastaldo et al. 2005; DiMichele and Gastaldo 2008).

Second, the amount of time averaging of nonmarine fossil assemblages is expected to increase over the history a basin. Rapid aggradation during the early high-accommodation phase of a basin favors the rapid burial of remains in both channel and floodplain deposits, whereas in the later, slowly aggrading low-accommodation phase, remains accumulate over longer periods of time before burial. Fossil assemblages in facies forming under higher-accommodation conditions should have generally lower amounts of time-averaging than assemblages forming under low-accommodation conditions. The extensive cannibalization of sediments involved in the formation of multistory channel sandstones increases the amount of reworking and time-averaging of vertebrate assemblages (Bown and Kraus 1981; Badgley 1986; Aslan and Behrensmeyer 1996; Gastaldo et al. 2017). This cannibalization allows the renewed oxidation and destruction of plant remains (Gastaldo and Demko 2011). Moreover, because channels typically sample over larger areas than floodplain, pond, and lake deposits, channel assemblages can mix taxa from different environments (Badgley 1986; Behrensmeyer and Chapman 1993; Rogers 1993; Behrensmeyer et al. 2000), contributing to greater spatial averaging in low-accommodation settings. This trend of increasing time averaging corresponds with increasing paleosol maturity. Pedogenesis proceeds over time in places receiving low amounts of sediment. Such settings are also able to accumulate the remains of plants and animals over longer periods of time, and so fossil assemblages in mature paleosols may be more time averaged than assemblages in immature paleosols or unmodified sediments (Aslan and Behrensmeyer 1996). Prolonged exposure, however, does not favor the preservation of plant remains and contributes to the weathering and destruction of bone, so mature paleosols may not host abundant or well-preserved fossil accumulations.

Third, the apparent overall nonmarine diversity of a basin is expected to decline over its history. The facies favored under high-accommodation conditions early in the history of a basin tend to have high preservation potential for a wide variety of major fossil groups (Fig. 7; Behrensmeyer and Hook 1992). Consequently, overall diversity is expected to be relatively high early in the history of a basin. With the net decline in accommodation rates, the later history of a basin is increasingly dominated by facies that preserve fewer major fossil groups (Fig. 7). The result is that diversity is expected to be lower late in the history of a basin. Within individual major groups, most are also expected to show a decline in diversity owing to the shift in the dominant facies and taphonomic gradients caused by the decline in rates of accommodation. For example, the long-term decline in the quality of plant and bone preservation will create an apparent decline of diversity within each of these groups. Aquatic mollusks might show the opposite pattern of increasing diversity through time as the proportion of channel deposits increases.

Different sedimentary basins will predictably show these trends to differing degrees. For example, the decline in subsidence rates over time is greater for foreland and extensional basins than for passive-margin and intracratonic basins (Fig. 14). In addition, because the nonmarine portion of passive-margin and intracratonic basins is always on the low-accommodation side of the basin (Fig. 11), accommodation rates there remain consistently low. As a result, trends in taphonomy and the representation of major groups will be more pronounced in foreland and extensional basins. In contrast, passive-margin and intracratonic basins are expected to have a persistently lower quality of preservation, except where eustatic rises in coastal basins temporarily create high-accommodation conditions.

Among basin types, there is a tradeoff between the quality of fossil preservation and basin size and longevity. Passive-margin and intracratonic basins characterized by persistently low rates of accommodation will preserve fossil records characterized by consistently relatively poor preservation quality, despite their large areas and long lifespans. In contrast, foreland and extensional basins will preserve extensive records of high-accommodation conditions with well-preserved fossils, but over relatively smaller areas and shorter spans of time.

7.2 Cycles in Taphonomic Mode and Preserved Biotas

The taphonomic mode and the occurrence of major groups of nonmarine fossils will also change predictably within depositional sequences (Fig. 21). Cyclical changes in accommodation and sedimentation have the potential to create atypical conditions within high- and low-accommodation settings. For example, high-accommodation settings will tend to be dominated by high-accommodation systems tracts, but they can preserve low-accommodation systems tracts during eustatic fall in coastal basins or when subsidence rates are reduced in any basin type. Similarly, low-accommodation settings have the potential to preserve high-accommodation systems tracts during pulses of subsidence or rapid eustatic rise, even though they might generally be dominated by low-accommodation systems tracts (Fig. 17). As a result, the relative proportions of preserved facies and the taphonomic characteristics of assemblages in the deposits will change cyclically within depositional sequences (Gastaldo et al. 2005; DiMichele and Gastaldo 2008). These shorter-term variations may mask long-term trends in taphonomy and preserved biotas, especially if the amplitude of the shorter-term changes in accommodation and sedimentation are large.

Systems tracts that form under rapid rates of accommodation (HAST and TST) will favor the development and preservation of floodplain, mire, and lake settings, all of which tend to have fossil assemblages characterized by generally good preservation, relatively low amounts of time averaging, and higher diversity (Fig. 21). Enhanced preservation in these settings also weights the fossil record towards wetland communities. Systems tracts that form under low or negative rates of accommodation (LAST, LST, HST, and especially FSST) will preserve facies containing fossil assemblages with generally poorer preservation quality, greater time averaging, and lower diversity (Fig. 21; Gastaldo et al. 2005). Thus, at the scale of depositional sequences, the quality of fossil preservation will systematically vary, especially so for the macrofossil record. These cyclical changes in taphonomic modes can also create gaps in the ranges of species and higher taxa. If the rates of change in accommodation are relatively slow, cycles in taphonomic mode and preserved biotas may be internally gradual and continuous. Where the rates of change in accommodation are high, there may be abrupt transitions in the preservation of major groups and their taphonomy. Such abrupt transitions are particularly likely at surfaces such as sequence boundaries, expansion surfaces, and in coastal basins, the updip equivalents of major marine flooding surfaces.

Recognizing the importance of stratigraphic architecture in controlling taphonomic mode at fine stratigraphic scales becomes particularly important during major biotic crises. For example, a loss of floral diversity near the

Permo–Triassic boundary in the Karoo Basin was initially attributed to widespread landscape denudation during the extinction (Ward et al. 2000). Subsequent study demonstrated not only that this change did not coincide with the mass extinction in the marine realm, but that the loss of diversity reflected changes in taphonomic mode controlled by stratigraphic architecture (Gastaldo et al. 2005; Gastaldo et al. 2020).

7.3 Fossil Concentrations at Predictable Stratigraphic Locations

Fossil concentrations are expected to occur at predictable locations, not just relative to facies, but within depositional sequences. Fossil concentrations include laterally continuous fossil-rich horizons (such as bonebeds, sensu Rogers et al. 2007), localized deposits with abundant fossils, and localized fossiliferous deposits occurring in close stratigraphic proximity. Of previous studies of stratigraphic paleobiology in nonmarine systems, by far the most attention has been given to the stratigraphic position of fossil concentrations, especially for vertebrates (e.g., Bown and Kraus 1981; Behrensmeyer 1987, 1988, 1991; Dodson 1987; Smith 1993; Badgley and Behrensmeyer 1995; Behrensmeyer et al. 2000; Eberth et al. 2001; Rogers and Kidwell 2000, 2007; Rogers et al. 2007, 2016; Rogers and Brady 2010; Loughney et al. 2011).

Fossil concentrations are controlled by the rate of hard-part input and preservation (R-hard-parts) and the rate of sedimentation (R-sediment; Kidwell 1986). Low net sedimentation can be achieved through omission, such as sediment starvation, dynamic bypassing, total passing, or through erosion (Kidwell 1986). In nonmarine systems, these conditions occur in stratigraphically predictable locations (Behrensmeyer 1991; Behrensmeyer et al. 2000; Rogers and Kidwell 2000, 2007). Some of these locations are related to depositional environments, such as the bases of river channels, the troughs of aqueous dunes, coals, and mature paleosols. Others are tied to sequence-stratigraphic architecture, such as the base of incised valley systems, exceptionally long-lived paleosols on interfluves (although prolonged weathering may destroy most bone), and flooding surfaces in lacustrine systems (Kidwell 1991). Still others reflect a combination of facies and sequence-stratigraphic architecture, in that systems tracts govern the relative frequency of facies, in turn controlling the occurrence of fossil concentrations.

A few studies have explored the importance of stratigraphic architecture for nonmarine vertebrates (notably Rogers and Kidwell 2000; Eberth et al. 2001; Loughney and Badgley 2017). For example, vertebrate concentrations are associated with several types of discontinuity surfaces in the Cretaceous Judith River Formation of the western interior, USA (Rogers and Kidwell

2000). There, an updip extension of a third-order marine transgressive surface lacks any bone concentrations, although it marks a change to increasing vertebrate-fossil abundance from underlying LAST to overlying HAST (Rogers et al. 2016). Similarly, a set of multistory fluvial sandstones associated with a sequence boundary also lacks vertebrate concentrations, which is interpreted to represent the prolonged destruction of bone in channel-dominated environments (Rogers et al. 2016). Increased vertebrate abundance in a HAST was also noted in Jurassic–Cretaceous Junggar Basin strata of China (Eberth et al. 2001), consistent with the greater abundance of vertebrates in floodplain deposits compared with channel deposits in the Triassic Ischigualasto Formation of Argentina (Colombi et al. 2013). In some cases, the reworking of preexisting sources of skeletal material in floodplain deposits may contribute more to the formation of bone accumulations than the magnitude of an erosional vacuity or the duration of a hiatus (Rogers and Kidwell 2000). In the Miocene Barstow Formation of California, bone concentrations show a strong association with particular facies, being abundant in abandoned-channel and spring-fed wetland deposits, somewhat less abundant in floodplain deposits, and rare in alluvial-fan (Fig. 13, A) and lake-margin deposits (Loughney and Badgley 2017). In the Barstow Formation, progressive upsection increase in the abundance of fossil concentrations reflects the proportional changes in these facies controlled by the long-term decline in rates of accommodation.

Fossil-plant accumulations also show a strong relationship to sequence-stratigraphic architecture (Gastaldo et al. 2005; Gastaldo and Demko 2011). In the Karoo Basin of South Africa, fluvial architecture is a primary control on plant fossils below and above the vertebrate-defined Permian–Triassic boundary, as identified in the early twenty-first century. Change in fluvial style from reportedly high-sinuosity to low-sinuosity systems in the latest Permian coincided with a decrease in macrofloral preservation and diversity in sandy braid-plain facies (Gastaldo et al. 2005). Throughout a depositional sequence, the position of the water table relative to the land surface governs rates of plant-fossil oxidation and, therefore, plant-fossil preservation. Periods of rapid aggradation (e.g., TST, HAST) have the greatest potential for preserving concentrations of plant fossils, compared with times of degradation of the landscape during base-level fall, such as in the FSST (DiMichele and Gastaldo 2008; Gastaldo and Demko 2011). Because base-level fall also fosters oxidation of plant remains in underlying strata, plant concentrations are less likely to occur in the upper HST and LAST (Gastaldo and Demko 2011).

Coal beds are also stratigraphically important fossil concentrations. Although the preservation of plant remains is often poor in coal (except where coal balls occur; e.g., Scott and Rex 1985; Siewers and Phillips 2015), coal is composed nearly entirely of organic remains and is itself a fossil concentration. Numerous resource-oriented studies have established the strong link between coal and sequence-stratigraphic architecture (Ryer 1984; Diessel 1988; Flint et al. 1995; Banerjee et al. 1996; Bohacs and Suter 1997; Hampson et al. 1999; Diessel et al. 2000; Holz et al. 2002; Wadsworth et al. 2002, 2010; Gibling et al. 2004). Specifically, rheotrophic mires are developed best during times of relative sea-level rise and consequent raising of the coastal-plain water table (Bohacs and Suter 1997). Of critical importance is the rate of increase in accommodation relative to the peat-production rate (Bohacs and Suter 1997; Wadsworth et al. 2002). If this ratio is too low, coals are thin and discontinuous, owing to insufficient accommodation. When these rates are balanced, continuously increasing accommodation sustains the peat-production rate, resulting in thick coals. If the rate of accommodation is too high, coal may initially accumulate but is subsequently drowned by lakes or the sea. In many cases, coal accumulation is closely tied to hiatal surfaces that correspond to marine flooding surfaces (Flint et al. 1995; Banerjee et al. 1996; Wadsworth et al. 2002, 2010; Gibling et al. 2004). Coals can also form in regressive intervals through the formation of raised mires in which the water table is perched above the regional water table (Diessel 1988; Banerjee et al. 1996; Gastaldo 2010). Even so, the ultimate preservation of the coal still requires the availability of sufficient accommodation (Diessel et al. 2000). In many cases, coals have a compound history with an earlier transgressive phase and a later regressive phase (Banerjee et al. 1996). Within depositional sequences, thick coals are common in the late LST to early TST and the late TST to early HST (Bohacs and Suter 1997). They can also form in other parts of a depositional sequence when the rates of accommodation and peat production are balanced.

7.4 Undetected Elevation Gradients in Community Composition

Sedimentary basins also experience long-term and cyclical changes in elevation owing to changes in the fluvial profile during degradation, progradation, aggradation, and retrogradation (Fig. 6). Although gradients in community composition correlated with elevation have been extensively documented by modern ecologists, examples from the fossil record are scarce. We hypothesize that these gradients are common but have lain undetected.

Testing for the occurrence of elevation-correlated gradients in community composition in the fossil record could be done in two ways, and both focus on detecting the climatic basis of elevation gradients. The first approach could be employed more readily in coastal basins. In such basins, nonbiostratigraphic means of correlation, such as magnetic reversals or distinctive ash beds, could be used to constrain a relatively narrow stratigraphic interval. Fossil collections could be made along a regional lateral transect through that interval, from areas adjacent to the coast and progressively distal to it. Community composition in these collections would be analyzed with ordination methods (Jongman et al. 1995; McCune and Grace 2002; Patzkowsky and Holland 2012; Borcard et al. 2018). Sample scores could be compared with distance from shore to test for a correlation. If a correlation were found, this would necessarily imply an elevation gradient because the sampling design is perpendicular to the coast. The ordination axis with the strongest correlation with distance from the coast correlates would also indicate the relative importance of the elevation gradient for community structure. For example, a correlation with ordination axis 1 would indicate that elevation is the primary ecological gradient. Qualitative tests for elevation gradients could also be conducted by comparing the taxonomic composition of basin-margin and basin-center assemblages that are taphonomically similar (isotaphonomic; Rogers et al. 2017) from stratigraphically restricted intervals in inland-basin deposits (Gunnell and Bartels 2001).

The second approach is more indirect, but it could be used in both coastal and inland basins. In this case, a correlation of community composition with temperature and precipitation could be tested, using isotopic or plant-morphology climate proxies (Wilf 1997; Peppe et al. 2011). Community composition could be described with ordination methods, and sample scores compared with quantitative proxies to test for a correlation. The difficulty in this approach is that the correlation with climate proxies might not necessarily indicate any relationship with elevation but may instead reflect temporal climate change or spatial climate variability uncorrelated with elevation. Separation of samples into those known to lie close to the shore and those that were more inland would allow any elevational component to this correlation to be isolated.

In both approaches, it would be crucial to minimize the distance sampled parallel to shore or lines of equal elevation to avoid the contribution of those spatial gradients. For example, in a study of the Atlantic Coastal Plain of the USA, one would want to minimize the latitudinal variation of the samples. If not, the ordination patterns would likely be dominated by the strong climate variations that exist from Florida to the mid-Atlantic.

7.5 Trends in Community Composition

Owing to net lengthening of the fluvial profile over the history of a sedimentary basin, elevation at any location should show a net increase (Fig. 6). This should be reflected in a net stratigraphic trend towards higher-elevation communities of plants, invertebrates, and vertebrates. Net lengthening of the fluvial profile arises largely from the long-term decline in basin subsidence rates, often paralleled by increasing sediment supply as source areas are eroded (e.g., Heller et al. 1988).

Because ecological gradients in nonmarine systems are commonly coupled to the climate-controlled elevation gradient, stratigraphic columns that span the history of a basin will show a net change from sampling relatively low-elevation settings to higher-elevation settings. Substantial evolution of species is likely to take place over the tens of millions (foreland, extensional) to hundreds of millions of years (passive margin, intracratonic) in which a basin is active. As a result, vertical stratigraphic trends in biotas will reflect a combination of evolutionary patterns, immigrations, extirpations, and increasing elevation. Any superimposed climatic trends will complicate the interpretation of the vertical stratigraphic change in biotas, as will any cyclical changes in base level. Identifying the contributions of all of these sources of biotic change will require carefully designed sampling strategies to isolate any one of them.

Apparently, few if any previous studies have interpreted such long-term patterns of community change in nonmarine biotas as reflecting changes in elevation. Even so, Carboniferous plants of the eastern USA may provide an example of a change in community composition (Falcon-Lang et al. 2011, 2018) that is consistent with the expected change from relatively low-elevation, coast-proximal settings to higher-elevation, coast-distal settings. In the Middle to Upper Pennsylvanian, a stepwise loss of lepidodendrids and other lycopsids along with the appearance of several types of coniferopsids is interpreted to reflect drier conditions. Similar patterns have been reported from other basins globally, and although this drying has been attributed to climate change (DiMichele et al. 2006), it may also reflect the drying that would be expected as areas that were once coastal become increasingly inland. Moreover, as basins fill with sediment and the shoreline regresses, the increasing land area promotes drier and more continental climates. Although much of this floral change occurs at the Middle/Upper Pennsylvanian (Desmoinesian/Missourian) boundary (Falcon-Lang et al. 2011), the pattern of biotic change is progressive and stepwise (DiMichele et al. 1985, 2010; DiMichele and Aronson 1992; Falcon-Lang et al. 2018). This pattern is consistent with ongoing progradation of coastal plains, accompanied by an increase in elevation and distance from the coast.

7.6 Cycles in Community Composition

As the vertical and lateral position of the shoreline in coastal basins changes through a depositional sequence, the fluvial profile is forced to adjust (Holbrook et al. 2006), driving cyclical changes in elevation. In turn, these changes are predicted to create stratigraphic cycles in community composition that correspond with depositional sequences.

Because shoreline trajectories map onto the four systems tracts used in marine and coastal sequence stratigraphy (LST, TST, HST, FSST; Table 1; Catuneanu 2006), changes in elevation along the coastal plain can be predicted within a depositional sequence. The LST will be characterized initially by a gain in elevation along the fluvial profile, but as shoreline trajectories become increasingly aggradational, upstream areas will begin to lose elevation. The TST and early HST will be characterized by elevation loss along the entire fluvial profile. As the HST becomes increasingly progradational, downstream areas will begin to gain elevation, which will spread landwards through the HST. In the latest HST and FSST, all locations will gain elevation. In inland basins, similar changes in elevation will track cycles of base level, driven by the elevation of an axial drainage system or lake level (Holbrook et al. 2006).

It is likely that the strength of these elevation-related community changes will depend not only on the magnitude of base-level change, but also on the position along the elevation gradient. For example, owing to the strong temperature and precipitation gradients immediately adjacent to the coast, as well as the intrusion of saltwater into rivers and the effects of salt spray, changes in community composition may be more pronounced closer to the coast than farther inland.

As the cycle periodicity of changes in accommodation and sedimentation becomes shorter, speciation and extinction are likely to play a lesser role in community composition. As such, short-period cyclicity (e.g., <1 Myr) in community composition may be a good guide to how the elevation gradient is expressed in any particular biota (e.g., DiMichele et al. 2007; Falcon-Lang et al. 2009). Regional climate change could also produce the same pattern, and testing would be needed to distinguish between the two. Such short-term changes in community composition may guide how to interpret long-term patterns (e.g., >10 Myr) of community change, where evolution and climate change are increasingly likely to contribute to the overall pattern of community change.

The amount of change in community composition will also depend on the rates of accommodation change in each systems tract. For example, the rate of sedimentation exceeds the rate of accommodation in the HST and LST (Fig. 17), leading to progradation of the coastal plain and increasing elevation. However, rates of accommodation decline through the HST but increase through the LST (Fig. 17), which causes more rapid progradation of the coastal plain in the HST than the LST (Fig. 18). As a result, the HST would be expected to show a strong pattern of elevation-related biotic change, whereas the LST might display little or no change. Because rates of accommodation peak in the middle of the TST (Fig. 17), the rate of decrease in elevation and rates of community change should likewise peak towards the middle of the TST.

Some of the most promising evidence of cyclical elevation-related changes in community composition again comes from Carboniferous plants of the eastern USA (Falcon-Lang et al. 2009; Falcon-Lang and DiMichele 2010). These records display an alternation within depositional sequences between humid-adapted lowland floras dominated by lycopsids and more seasonally dry-adapted "upland" (higher-elevation) coniferous floras. Preservation of the lowland floras is overwhelmingly more common, owing to the need for an elevated water table to preserve plants (Fig. 10). The stratigraphic pattern of these floral changes is likely complicated considerably by sequence-stratigraphic architecture and the selective preservation of depositional environments. Although these floral changes are interpreted as reflecting changes in global climate from humid interglacials to seasonally dry glacials (Falcon-Lang 2004), they are also consistent with the expected local changes in physiographic setting during cycles of sea level. Indeed, spectacular preservation of a Pennsylvanian drowned forest shows that these lower-elevation and higher-elevation floras were coeval and persistent, not simply different communities that existed under distinct climatic regimes (DiMichele et al. 2007; Tabor et al. 2013). That these were coeval communities is also supported by palynofloras (Looy and Hotton 2014; Looy et al. 2014). For plants, incised valleys are critically important archives of higher-elevation communities. Even in these settings, however, the fossil record is dominated by the preservation of relatively low-elevation environments in which the water table was relatively high, as opposed to settings on interfluves (Demko et al. 1998).

7.7 Abrupt Community Change across Sequence Boundaries

Abrupt change in nonmarine biotas is expected across sequence boundaries because hiatuses provide time for turnover in fossil assemblages. Any species that goes extinct or that is regionally extirpated during the time of the hiatus will have its last occurrence below the sequence boundary. Similarly, any species that originates or immigrates during the hiatus will have its first occurrence above the sequence boundary. If the sequence boundary is either underlain or overlain by marine facies, the effective hiatus in the nonmarine fossil record increases, allowing for even more turnover to accumulate and generating an even more abrupt change in biotas (Gastaldo et al. 2009).

Because the formation of an unconformity requires erosion and degradation of the landscape (Blum et al. 2013), nonmarine facies overlying and underlying a sequence boundary will commonly represent different elevations and sedimentary environments. As a result, nonmarine biotas from different elevations will likely be superimposed at a sequence boundary, adding to the apparent turnover and increasing the number of first and last occurrences at the sequence boundary, similar to what occurs at sequence boundaries in marine systems (Holland 1995).

The proximity of first and last occurrences to the sequence boundary will depend on the abundance of the species, its preferred habitat, the facies above and below the boundary, and the eurytopy of the species, following reasoning similar to marine habitats (Holland and Patzkowsky 2002). Higher species abundance increases the probability of occurrence of a species, causing first and last occurrences to lie close to the sequence boundary. Similarly, species whose preferred habitat is represented by facies immediately below a sequence boundary will have their last occurrence close to the sequence boundary. A similar logic applies to the first occurrence of species whose preferred habitat is recorded by the facies overlying the sequence boundary. Finally, because eurytopic species are broadly distributed among facies, it is more likely that their first and last occurrences will lie close to the sequence boundary.

7.8 Abrupt Community Change across Updip Flooding Surfaces

Abrupt changes in nonmarine biotas are expected across the updip extension of marine flooding surfaces in coastal basins (Fig. 18; Table 1). In coastal nonmarine areas, a rise in relative sea level also elevates the water table. If physicochemical conditions permit (e.g., Gastaldo 2010), this rise leads to the formation and preservation of coal and organic-rich sediment on the lower coastal plain at the same time that the marine flooding surface forms downdip (Shanley and McCabe 1994; Aitken and Flint 1995; Bohacs and Suter 1997; Wadsworth et al. 2002, 2010; Gastaldo and Demko 2011). Coal formation coincides with a landward translation of the shoreline and a change in the topographic profile of the coastal plain. This change should be manifested in the rock record as lower coastal-plain (i.e., lower-elevation) communities superimposed on more inland coastal-plain (higher-elevation) communities.

Owing to the peak in accommodation rates in the middle of the TST (Fig. 17), the landward rate of shoreline displacement during the formation of a flooding surface also peaks in the middle of the TST. In the marine realm, the effect is that flooding surfaces in the TST record the greatest amount of facies change in a depositional sequence. In the nonmarine realm, this is also true on the lower coastal plain. As a result, changes in coastal biotas associated with coals will be more abrupt in the TST than in the HST or LST. Consequently, the overall pattern would be a punctuated trend towards lower-elevation biotas of the lower coastal plain in the TST, followed by a more gradual and less punctuated shift towards higher-elevation, more inland biotas during the HST. Similarly, low rates of accommodation in the LST would favor a more gradual change in community composition than a punctuated one.

In inland basins, surfaces equivalent to marine flooding surfaces are mostly absent, except where large lake systems display lacustrine flooding surfaces analogous to marine flooding surfaces. An increase in base level accompanying the rapid growth of a lake would cause the updip translation in depositional environments and shift in the fluvial profile (similar to the retrogradation and aggradation case in Fig. 6; Holbrook et al. 2006). In areas not flooded by the lake, this would be recorded as an abrupt shift to relatively higher-elevation biotas.

Fluvial aggradational cycles (FACs) have been described based on changes in soil maturity and drainage characteristics (Atchley et al. 2004). Because their bounding surfaces are caused by local depositional dynamics, they are more similar to the aggradational-stasis cycles of Gastaldo and Demko (2011) rather than large-scale changes in fluvial gradients. As such, the bounding surfaces of FACs are not predicted to record substantial change in nonmarine biotas.

8 Closing Comments

The fossil record is undeniably complex, and we are not suggesting that stratigraphic architecture is the only control on the nonmarine fossil record. We emphasize, however, that some aspects of the nonmarine fossil record likely do arise from stratigraphic architecture, and we present here eight broad sets of predictions of how stratigraphic architecture may be manifested in the nonmarine fossil record. These predictions are based on ecological, taphonomic, and stratigraphic observations, such as the existence of climate-correlated gradients in species composition, environmental patterns in taphonomy, the long-term decline in subsidence rates in sedimentary basins, and the controls on nonmarine stratigraphic architecture. Where these underlying patterns diverge from what we describe here, they may lead to alternative predictions, which can likewise be tested. For example, basins that do not undergo a long-term change in subsidence rates are not expected to show long-term trends in taphonomy and elevation-correlated changes in species composition.

We fully acknowledge that other controls on the nonmarine fossil and stratigraphic records exist, such as climate variability, spatial and temporal ecological variation unrelated to elevation, and autogenic dynamics. It is essential to recognize, though, that simply the processes of stratigraphic accumulation can create patterns in the fossil record, and that these patterns require no special biological interpretation.

Which sets of controls on the nonmarine fossil record are the most important is an open question, and we hope that this work inspires a rich investigation of these drivers, much as has happened for marine systems. Almost certainly there is no single answer to which sets of controls are most important, and scale will be crucial (Bennington et al. 2009). For example, the elevation-correlated gradient in ecological composition may matter most where elevation changes are the greatest, such as over the history of a basin or during high-amplitude sea-level change in coastal basins, or where communities are most sensitive to small changes in elevation, such as areas close to a coast. In other situations, elevation-related community change may be overwhelmed by other sources of community variation. These are among the many interesting questions raised by these hypotheses, and we look forward to the studies to come.

9 Conclusions

1. The nonmarine fossil record is shaped by the stratigraphic record in which it is preserved. Nonmarine strata record progressive and cyclical changes in their architecture and in the elevation-correlated gradients they preserve. These trends and cycles control the taphonomy, stratigraphic position, and taxonomic composition of nonmarine fossil assemblages.
2. Elevation-correlated gradients in community composition are significant because ecologists have demonstrated that modern nonmarine biotas are primarily distributed along gradients in moisture and temperature that are commonly correlated with elevation and distance from the coast. These gradients are hypothesized to be common but widely undetected in the nonmarine fossil record.
3. Facies and stratigraphic architecture in nonmarine basins reflect the long-term decline in subsidence rates observed in most sedimentary basins. This decline is predicted to impart long-term trends in the composition and preservation quality of the nonmarine fossil record in individual basins. In large part, these patterns arise from systematic changes in the proportions of various nonmarine facies preserved under high-accommodation and low-accommodation conditions. Over their history, the nonmarine fossil records of individual basins are expected to show decreasing quality of preservation, increasing amounts of time averaging, and declining diversity. Owing to the greater range of subsidence rates over their history, particularly in their

nonmarine areas, foreland and extensional basins are predicted to display these patterns more strongly than passive-margin and intracratonic basins.

4. Superimposed on the long-term decline in rates of accommodation will be cycles of accommodation that generate depositional sequences. Their systems tracts will add systematic variations in the types of depositional settings preserved and will create cyclical patterns in quality of preservation, time averaging, and apparent diversity.

5. Basins with nonmarine deposits are predicted to record a net trend from low-elevation to high-elevation communities through their history, owing to the lengthening of fluvial profiles as the basin fills with sediment. Depositional sequences are also hypothesized to record cyclical changes in the elevation indicated by fossil communities. Sequence boundaries are expected to record abrupt changes in community composition in all basins, as are marine flooding surfaces in coastal basins. Methods are proposed for testing the existence of elevation-related gradients in community composition.

References

Aitken, J., and S. Flint. 1995. The application of high-resolution stratigraphy to fluvial systems: a case study from the Upper Carboniferous Breathitt Group, eastern Kentucky, USA. Sedimentology 42:3–30.

Al-Aklabi, A., A. W. Al-Khulaidi, A. Hussain, and N. Al-Sagheer. 2016. Main vegetation types and plant species diversity along an altitudinal gradient of Al Baha region, Saudi Arabia. Saudi Journal of Biological Sciences 23:687–97.

Allen, P. A., and J. R. Allen. 2005. Basin Analysis: Principles and Applications. New York, Wiley-Blackwell.

Amante, C. 2009. ETOPO1 1 arc-minute global relief model: procedures, data sources, and analysis. U. S. Department of Commerce, National Oceanic and Atmospheric Administration, National Geophysical Data Center, Marine Geology and Geophysics Division, Boulder, Colorado.

Angevine, C. L., P. L. Heller, and C. Paola. 1990. Quantitative sedimentary basin modeling. American Association of Petroleum Geologists Shortcourse Note Series 32.

Armitage, J. J., and P. A. Allen. 2010. Cratonic basins and the long-term subsidence history of continental interiors. Journal of the Geological Society 167:61–70.

Arnott, R. W. C., B. A. Zaitlin, and D. J. Potocki. 2002. Stratigraphic response to sedimentation in a net-accommodation-limited setting, Lower Cretaceous Basal Quartz, south-central Alberta. Bulletin of Canadian Petroleum Geology 50:92–104.

Aslan, A., and A. K. Behrensmeyer. 1996. Taphonomy and time resolution of bone assemblages in a contemporary fluvial system: the East Fork River, Wyoming. Palaios 11:411–21.

Atchley, S. C., L. C. Nordt, and S. I. Dworkin. 2004. Eustatic control on alluvial sequence stratigraphy: a possible example from the Cretaceous–Tertiary transition of the Tornillo Basin, Big Bend National Park, west Texas, U.S. A.Journal of Sedimentary Research 74:391–404.

Austin, M. P., R. B. Cunningham, and P. M. Fleming. 1984. New approaches to direct gradient analysis using environmental scalars and statistical curve-fitting procedures. Plant Ecology 55:11–27.

Badgley, C. 1986. Taphonomy of mammalian fossil remains from Siwalik rocks of Pakistan. Paleobiology 12:119–42.

Badgley, C., and A. K. Behrensmeyer. 1995. Preservational, paleoecological and evolutionary patterns in the Paleogene of Wyoming–Montana and the

Neogene of Pakistan. Palaeogeography, Palaeoclimatology, Palaeoecology 115:319–40.

Badgley, C., and D. L. Fox. 2000. Ecological biogeography of North American mammals: species density and ecological structure in relation to environmental gradients. Journal of Biogeography 27:1437–67.

Badgley, C., T. M. Smiley, R. Terry, E. B. Davis et al. 2017. Biodiversity and topographic complexity: modern and geohistorical perspectives. Trends in Ecology and Evolution 32:211–26.

Banerjee, J., W. Kalkreuth, and E. Davies. 1996. Coal seam splits and transgressive–regressive coal couplets: a key to stratigraphy of high-frequency sequences. Geology 24:1001–4.

Behrensmeyer, A. K. 1978. Taphonomic and ecologic information from bone weathering. Paleobiology 4:150–62.

1982. Time resolution in fluvial vertebrate assemblages. Paleobiology 8:211–27.

1987. Miocene fluvial facies and vertebrate taphonomy in northern Pakistan. In:F. G. Ethridge, R. M. Flores, and M. D. Harvey, eds. Recent Developments in Fluvial Sedimentology. Society of Economic Paleontologists and Mineralogists Special Publication 39:169–76.

1988. Vertebrate preservation in fluvial channels. Palaeogeography, Palaeoclimatology, Palaeoecology 63:183–99.

1991. Terrestrial vertebrate accumulations. In: P. A. Allison and D. E. G. Briggs, eds. Taphonomy: Releasing the Data Locked in the Fossil Record. Plenum Press, New York: 291–335

Behrensmeyer, A. K., and R. E. Chapman. 1993. Models and simulations of time-averaging in terrestrial vertebrate accumulations. Short Courses in Paleontology 6:125–49.

Behrensmeyer, A. K., and R. W. Hook. 1992. Paleoenvironmental contexts and taphonomic modes. In: A. K. Behrensmeyer, J. D. Damuth, W. A. DiMichele, R. Potts, H. D. Sues, and S. L. Wing, eds. Terrestrial Ecosystems through Time. University of Chicago Press, Chicago: 15–136

Behrensmeyer, A. K., and S. M. Kidwell. 1985. Taphonomy's contributions to paleobiology. Paleobiology 11:105–19.

Behrensmeyer, A. K., B. J. Willis, and J. Quade. 1995. Floodplains and paleosols of Pakistan Neogene and Wyoming Paleogene deposits: a comparative study. Palaeogeography, Palaeoclimatology, Palaeoecology 115:37–60.

Behrensmeyer, A. K., S. M. Kidwell, and R. A. Gastaldo. 2000. Taphonomy and paleobiology. Deep Time: Paleobiology's Perspective 26:103–47.

Bennington, J. B., W. A. DiMichele, C. Badgley, et al. 2009. Critical issues of scale in paleoecology. Palaios 24:1–4.

Blum, M. D., J. Martin, K. Milliken, and M. Garvin. 2013. Paleovalley systems: insights from Quaternary analogs and experiments. Earth-Science Reviews 116:128–69.

Bohacs, K., and J. Suter. 1997. Sequence stratigraphic distribution of coaly rocks: fundamental controls and paralic examples. American Association of Petroleum Geologists Bulletin 81:1612–39.

Borcard, D., F. Gillet, and P. Legendre. 2018. Numerical Ecology with R, second edition. Springer, Cham, Switzerland.

Bown, T. M., and M. J. Kraus. 1981. Vertebrate fossil-bearing paleosol units (Willwood Formation, Lower Eocene, northwest Wyoming, U.S.A.): implications for taphonomy, biostratigraphy, and assemblage analysis. Palaeogeography, Palaeoclimatology, Palaeoecology 34:31–56.

1987. Integration of channel and floodplain suites, I. developmental sequence and lateral relations of alluvial paleosols. Journal of Sedimentary Petrology 57:587–601.

Boyd, R., C. Diessel, J. Wadsworth, D. Leckie, and B. A. Zaitlin. 2000. Developing a model for non-marine sequence stratigraphy—application to the Western Canada Sedimentary Basin. GeoCanada 2000—The Millennium Geosciences Summit Conference Abstract.

Bridge, J. S. 2003. Rivers and Floodplains: Forms, Processes, and Sedimentary Record. Blackwell Science, Oxford.

Bryant, M., P. Falk, and C. Paola. 1995. Experimental study of avulsion frequency and rate of deposition. Geology 23:365–68.

Burnham, R. J. 1988. Paleoecological approaches to analyzing stratigraphic sequences. In: W. A. DiMichele and S. L. Wing, eds. Methods and Applications of Plant Paleoecology. Paleontological Society Special Publication 3:105–25.

Calder, J. H., M. R. Gibling, A. C. Scott, S. J. Davies, and B. L. Hebert. 2006. A fossil lycopsid forest succession in the classic Joggins section of Nova Scotia: paleoecology of a disturbance-prone Pennsylvanian wetland. In: W. A. DiMichele and S. F. Greb, eds. Wetlands through Time. Geological Society of America Special Paper 399:169–95.

Campo, B., A. Amorosi, and L. Bruno. 2016. Contrasting alluvial architecture of Late Pleistocene and Holocene deposits along a 120-km transect from the central Po Plain (northern Italy). Sedimentary Geology 341:265–75.

Catuneanu, O. 2006. Principles of Sequence Stratigraphy. Elsevier, Amsterdam.

Cecil, C. B., and F. T. Dulong. 2003. Precipitation models for sediment supply in warm climates. In: C. B. Cecil and N. T. Edgar, eds. Climate controls on stratigraphy. SEPM Special Publication 77:21–7.

Chamberlin, E. P., and E. A. Hajek. 2015. Interpreting paleo-avulsion dynamics from multistory sand bodies. Journal of Sedimentary Research 85:82–94.

Colombi, C. E., R. R. Rogers, and O. A. Alcober. 2013. Vertebrate taphonomy of the Ischigualasto Formation. Society for Vertebrate Paleontology Memoir 12:31–50.

Currie, B. S. 1997. Sequence stratigraphy of nonmarine Jurassic–Cretaceous rocks, central Cordilleran foreland-basin system. Geological Society of America Bulletin 109:1206–22.

Dail, M. R., J. R. Hill, and R. D. Miller. 2013. The Virginia coastal plain macroinvertebrate index. Virginia Department of Environmental Quality Bulletin WQA/2013–002.

Dale, E. E., Jr., S. Ware, and B. Waitman. 2007. Ordination and classification of bottomland forests in the lower Mississippi alluvial plain. Castanea 72:105–15.

Danise S., M. E. Clémence, G. D. Price, D. P. Murphy, J. Gómez, and R. J. Twitchett. 2019. Stratigraphic and environmental control on marine benthic community change through the Early Toarcian extinction event (Iberian Range, Spain). Palaeogeography, Palaeoclimatology, Palaeoecology 524:183–200.

DeCelles, P. G., and K. A. Giles. 1996. Foreland basin systems. Basin Research 8:105–23.

Demko, T. M., R. F. Dubiel, and J. T. Parrish. 1998. Plant taphonomy in incised valleys: implications for interpreting paleoclimate from fossil plants. Geology 26:1119–22.

Diessel, C. F. K. 1988. Sequence stratigraphy applied to coal seams: two case histories. Society of Economic Paleontologists and Mineralogists Special Publication In: K. W. Shanley and P. J. McCabe, eds. Relative Role of Eustasy, Climate, and Tectonism in Continental Rocks. Society of Economic Paleontologists and Mineralogists Special Publication 59:151–73

Diessel, C., R. Boyd, J. Wadsworth, D. Leckie, and G. Chalmers. 2000. On balanced and unbalanced accommodation/peat-accumulation ratios in the Cretaceous coals from Gates Formation, western Canada, and their sequence-stratigraphic significance. International Journal of Coal Geology 43:143–86.

DiMichele, W. A., and R. B. Aronson. 1992. The Pennsylvanian–Permian vegetational transition: a terrestrial analogue to the onshore–offshore hypothesis. Evolution 46:807–24.

DiMichele, W. A., and R. A. Gastaldo. 2008. Plant paleoecology in deep time. Annals of the Missouri Botanical Gardens 95:144–98.

DiMichele, W. A., and T. L. Phillips. 1994. Paleobotanical and paleoecological constraints on models of peat formation in the Late Carboniferous of Euramerica. Palaeogeography, Palaeoclimatology, Palaeoecology 106:39–90.

DiMichele, W. A., T. L. Phillips, and R. A. Peppers. 1985. The influence of climate and depositional environment on the distribution and evolution of Pennsylvanian coal-swamp plants. In B. H. Tiffney, ed. Geological Factors and the Evolution of Plants. Yale University Press, New Haven: 223–56.

DiMichele, W. A., N. J. Tabor, D. S. Chaney, and W. J. Nelson. 2006. From wetlands to wet spots: environmental tracking and the fate of Carboniferous elements in Early Permian tropical floras. In: W. A. DiMichele and S. F. Greb, eds. Wetlands through Time. Geological Society of America Special Paper 399:223–48.

DiMichele, W. A., H. J. Falcon-Lang, W. John Nelson, S. D. Elrick, and P. R. Ames. 2007. Ecological gradients within a Pennsylvanian mire forest. Geology 35:415–18.

DiMichele, W. A., C. B. Cecil, I. P. Montañez, and H. J. Falcon-Lang. 2010. Cyclic changes in Pennsylvanian paleoclimate and effects on floristic dynamics in tropical Pangaea. International Journal of Coal Geology 83:329–44.

Dodson, P. 1987. Sedimentology and taphonomy of the Oldman Formation (Campanian), Dinosaur Provincial Park, Alberta (Canada). Palaeogeography, Palaeoclimatology, Palaeoecology 10:21–74.

Dominici, S., and T. Kowalke. 2007. Depositional dynamics and the record of ecosystem stability: Early Eocene faunal gradients in the Pyrenean foreland, Spain. Palaios 22:268–84.

Eberth, D. A., D. B. Brinkman, P. J. Chen, et al. 2001. Sequence stratigraphy, paleoclimate patterns, and vertebrate fossil preservation in Jurassic–Cretaceous strata of the Junggar Basin, Xinjiang Autonomous Region, People's Republic of China. Canadian Journal of Earth Sciences 38:1627–44.

Edwards, L., J. Ambrose, and L. K. Kirkman, 2013. The Natural Communities of Georgia. University of Georgia, Athens, Georgia.

Falcon-Lang, H. J. 2004. Pennsylvanian tropical rain forests responded to glacial–interglacial rhythms. Geology 32:689–92.

Falcon-Lang, H. J., and W. A. DiMichele. 2010. What happened to the coal forests during Pennsylvanian glacial phases? Palaios 25:611–17.

Falcon-Lang, H. J., W. J. Nelson, S. Elrick, C. V. Looy, P. R. Ames, and W. A. DiMichele. 2009. Incised channel fills containing conifers indicate that seasonally dry vegetation dominated Pennsylvanian tropical lowlands. Geology 37:923–6.

Falcon-Lang, H. J., P. H. Heckel, W. A. DiMichele, et al. 2011. No major stratigraphic gap exists near the Middle–Upper Pennsylvanian (Desmoinesian–Missourian) boundary in North America. Palaios 26:125–39.

Falcon-Lang, H., W. John Nelson, P. H. Heckel, W. A. DiMichele, and S. D. Elrick. 2018. New insights on the stepwise collapse of the Carboniferous coal forests: evidence from cyclothems and coniferopsid tree stumps near the Desmoinesian–Missourian boundary in Peoria County, Illinois, USA. Palaeogeography, Palaeoclimatology, Palaeoecology 490:375–92.

Fielding, C. R. 1987. Coal depositional models for deltaic and alluvial plain sequences. Geology 15:661–4.

Fielding, C. R., P. J. Ashworth, J. I. Best, E. W. Prokocki, and G. H. Sambrook Smith. 2012. Tributary, distributary and other fluvial patterns: what *really* represents the norm in the continental rock record? Sedimentary Geology 261–2:15–32.

Flint, S., J. Aitken, and G. Hampson. 1995. Application of sequence stratigraphy to coal-bearing coastal plain successions: implications for the UK Coal Measures. In: M. K. G. Whateley and D. A. Spears, eds. European Coal Geology. Geological Society Special Publication 82, London: 1–16.

Gagnon, P., W. Michener, M. Freeman, and J. B. Box. 2006. Unionid habitat and assemblage composition in coastal plain tributaries of Flint River (Georgia). Southeastern Naturalist 5:31–52.

Gani, M. R., A. Ranson, D. B. Cross, G. J. Hampson, N. D. Gani, and H. Sahoo. 2015. Along-strike sequence stratigraphy across the Cretaceous shallow marine to coastal-plain transition, Wasatch Plateau, Utah, U. S. A. Sedimentary Geology 325:59–70.

Gastaldo, R. A. 1987. Confirmation of Carboniferous clastic swamp communities. Nature 326:869–71.

1988. A conspectus of phytotaphonomy. Paleontological Society Special Publication 3:14–28.

1992. Taphonomic considerations for plant evolutionary investigations. The Paleobotanist 41:211–23.

2010. Peat or no peat: why do the Rajang and Mahakam Deltas differ? International Journal of Coal Geology 83:162–72.

Gastaldo, R. A., and T. M. Demko. 2011. The relationship between continental landscape evolution and the plant-fossil record: long term hydrologic controls on preservation. In: P. A. Allison and D. J. Bottjer, eds.

Taphonomy, Second Edition: Processes and Bias Through Time. Topics in Geobiology 32:249–86.

Gastaldo, R. A., and J. R. Staub. 1999. A mechanism to explain the preservation of leaf litter lenses in coals derived from raised mires. Palaeogeography, Palaeoclimatology, Palaeoecology 149:1–14.

Gastaldo, R. A., D. P. Douglass, and S. M. McCarroll, 1987. Origin, characteristics and provenance of plant macrodetritus in a Holocene crevasse splay, Mobile delta, Alabama. Palaios 2:229–40.

Gastaldo, R. A., I. M. Stevanović-Walls, W. N. Ware, and S. F. Greb. 2004. Community heterogeneity of Early Pennsylvanian peat mires. Geology 32:693–4.

Gastaldo, R. A., R. Adendorff, M. K. Bamford, N. J. Labandeira, and H. J. Sims. 2005. Taphonomic trends of macrofloral assemblages across the Permian–Triassic boundary, Karoo Basin, South Africa. Palaios 20:478–97.

Gastaldo, R. A., E. Purkynova, Z. Simunek, and M. D. Schmitz. 2009. Ecological persistence in the Late Mississippian (Serpukhovian, Namurian A) megafloral record of the Upper Silesian Basin, Czech Republic. Palaios 24:336–50.

Gastaldo, R. A., J. Neveling, C. V. Looy, M. K. Bamford, S. L. Kamo, and J. W. Geissman. 2017. Paleontology of the Blaauwater 67 and 65 farms, South Africa: testing the *Daptocephalus/Lystrosaurus* biozone boundary in a stratigraphic framework. Palaios 32:349–66.

Gastaldo, R. A., S. L. Kamo, J. Neveling, J. Geissman, C. V. Looy, and A. M. Martini. 2020. The base of the *Lystrosaurus* Assemblage Zone, Karoo Basin, predates the end-Permian marine extinction. Nature Communications 11:1428.

Gawthorpe, R. L., and M. R. Leeder. 2000. Tectono-sedimentary evolution of active extensional basins. Basin Research 12:195–218.

Gibling, M. R. 2018. River systems and the Anthropocene: a Late Pleistocene and Holocene timeline for human influence. Quaternary 1:21–36.

Gibling, M. R., K. I. Saunders, N. E. Tibert, and J. A. White. 2004. Sequence sets, high-accommodation events, and the coal window in the Carboniferous Sydney Coalfield, Atlantic Canada. In: J. C. Pashin and R. A. Gastaldo, eds. Sequence Stratigraphy, Paleoclimate, and Tectonics of Coal-bearing Strata. AAPG Studies in Geology. American Association of Petroleum Geologists, Tulsa, Oklahoma, 51:169–97

Gibson, L. A., and N. L. McKenzie. 2009. Environmental associations of small ground-dwelling mammals in the Pilbara region, Western Australia. Records of the Western Australian Museum, Supplement 78:91–122.

Gordon, I., and P. L. Heller. 1993. Evaluating major controls on basinal stratigraphy, Pine Valley, Nevada: implications for syntectonic deposition. Geological Society of America Bulletin 105:47–55.

Gottesfeld, A. S. 1972. Paleoecology of the lower part of the Chinle Formation in the Petrified Forest. In: C. S. Breed, and W. J. Breed, eds. Investigations in the Triassic Chinle Formation. Museum of Northern Arizona Bulletin 47:59–73.

Greenwood, D. R. 1991. The taphonomy of plant macrofossils. In: S. K. Donovan, ed. The Processes of Fossilization. Columbia University Press, New York:141–69.

Grytnes, J. A., and O. R. Vetaas. 2002. Species richness and altitude: a comparison between null models and interpolated plant species richness along the Himalayan altitudinal gradient, Nepal. The American Naturalist 159:294–304.

Gulliford, A. R., S. S. Flint, and D. M. Hodgson. 2014. Testing applicability of models of distributive fluvial systems or trunk rivers in ephemeral systems: reconstructing 3-d fluvial architecture in the Beaufort Group, South Africa. Journal of Sedimentary Research 84:1147–69.

——— 2017. Crevasse splay processes and deposits in an ancient distributive fluvial system: the lower Beaufort Group, South Africa. Sedimentary Geology 358:1–18.

Gunnell, G. F., and W. S. Bartels. 2001. Basin margins, biodiversity, evolutionary innovation, and the origin of new taxa. In: G. F. Gunnell, ed. Eocene Biodiversity: Unusual Occurrences and Rarely Sampled Habitats. Kluwer Academic/Plenum, New York:403–32.

Hajek, E. A., and P. L. Heller. 2012. Flow-depth scaling in alluvial architecture and nonmarine sequence stratigraphy: example from the Castlegate Sandstone, Central Utah, U. S. A. Journal of Sedimentary Research 82:121–30.

Hajek, E. A., and M. A. Wolinsky. 2012. Simplified process modeling of river avulsion and alluvial architecture: connecting models and field data. Sedimentary Geology 257–60:1–30.

Hampson, G., H. Stollhofen, and S. Flint. 1999. A sequence stratigraphic model for the lower coal measures (Upper Carboniferous) of the Ruhr district, north-west Germany. Sedimentology 46:1199–231.

Harcombe, P. A., J. S. Glitzenstein, R. G. Know, S. L. Orzell, and E. L. Bridges. 1993. Vegetation of the longleaf pine region of the west Gulf Coastal Plain. In: S. M. Hermann, ed. Proceedings of the Tall Timbers Fire Ecology Conference, Tallahassee, Florida: 83–104.

Hartley, A. J., G. S. Weissmann, G. J. Nichols, and G. L. Warwick. 2010. Large distributive fluvial systems: characteristics, distribution, and controls on development. Journal of Sedimentary Research 80:167–83.

Heino, J., and J. Alahuhta. 2015. Elements of regional beetle faunas: faunal variation and compositional breakpoints along climate, land cover and geographical gradients. Journal of Animal Ecology 84:427–41.

Heller, P. L., C. L. Angevine, N. S. Winslow, and C. Paola. 1988. Two-phase stratigraphic model of foreland basin sequences. Geology 16:501–4.

Holbrook, J., R. W. Scott, and F. E. Oboh-Ikuenobe. 2006. Base-level buffers and buttresses: a model for upstream versus downstream control on fluvial geometry and architecture within sequences. Journal of Sedimentary Research 76:162–74.

Holland, S. M. 1995. The stratigraphic distribution of fossils. Paleobiology 21:92–109.

2000. The quality of the fossil record: a sequence stratigraphic perspective. Paleobiology 26:148–68.

2016. The non-uniformity of fossil preservation. Philosophical Transactions of the Royal Society of London, Series B, Biological Sciences 371:20150130–11.

2017. Presidential address: structure, not bias. Journal of Paleontology 91:1315–17.

Holland, S. M., and M. E. Patzkowsky. 2002. Stratigraphic variation in the timing of first and last occurrences. Palaios 17:134–46.

2007. Gradient ecology of a biotic invasion: biofacies of the type Cincinnatian series (Upper Ordovician), Cincinnati, Ohio region, USA. Palaios 22:392–407.

2015. The stratigraphy of mass extinction. Palaeontology 58:903–24.

Holz, W., and W. Kalkreuth. 2004. Sequence stratigraphy and coal petrology applied to the Early Permian coal-bearing Rio Bonito Formation, Paraná Basin, Brazil. In: J. C. Pashin and R. A. Gastaldo, eds. Sequence Stratigraphy, Paleoclimate, and Tectonics of Coal-bearing Strata. AAPG Studies in Geology. American Association of Petroleum Geologists, Tulsa, Oklahoma, 51:147–67.

Holz, M., W. Kalkreuth, and I. Banerjee. 2002. Sequence stratigraphy of paralic coal-bearing strata: an overview. International Journal of Coal Geology 48:147–79.

Hunt, D., and M. E. Tucker. 1992. Stranded parasequences and the forced regressive wedge systems tract: deposition during base-level fall. Sedimentary Geology 81:1–9.

Jackson, S. T., and J. T. Overpeck. 2000. Responses of plant populations and communities to environmental changes of the late Quaternary. Paleobiology 26 (S4):194–220.

Jackson, S. T., and J. W. Williams. 2004. Modern analogs in Quaternary paleoecology: here today, gone yesterday, gone tomorrow? Annual Review of Earth and Planetary Sciences 32:495–537.

Jongman, R. H. G., C. J. F. Ter Braak, and O. F. R. Van Tongeren, eds. 1995. Data Analysis in Community and Landscape Ecology. Cambridge University Press, Cambridge.

Jordan, T. E. 1995. Retroarc foreland and related basins. In: C. J. Busby and R. V. Ingersoll, eds. Tectonics of Sedimentary Basins. Blackwell, Oxford: 331–62.

Jordan, T. E., and P. B. Flemings. 1991. Large-scale stratigraphic architecture, eustatic variation, and unsteady tectonism: a theoretical evaluation. Journal of Geophysical Research 96:6681–99.

Kidwell, S. M. 1986. Models for fossil concentrations: paleobiologic implications. Paleobiology 12:6–24.

1991. The stratigraphy of shell concentrations. In: P. A. Allison and D. E. G. Briggs, eds. Taphonomy, Releasing the Data Locked in the Fossil Record. Plenum, New York:211–90.

1993. Taphonomic expressions of sedimentary hiatus: field observations on bioclastic concentrations and sequence anatomy in low, moderate and high subsidence settings. Geologische Rundschau 82:189–202.

Kraus, M. J. 1999. Paleosols in clastic sedimentary rocks: their geologic implications. Earth-Science Reviews 47:41–70.

Leeder, M. R. 1995. Continental rifts and proto-oceanic rift troughs. In: C. J. Busby and R. V. Ingersoll, eds. Tectonics of Sedimentary Basins. Blackwell, Oxford:119–48.

Leeder, M. R., and R. L. Gawthorpe. 1987. Sedimentary models for extensional tilt-block/half-graben basins. In: M. P. Coward, J. F. Dewey, and P. L. Hancock, eds. Continental Extensional Tectonics, Geological Society Special Publication 28:139–52.

Lehman, T. M. 2001. Late Cretaceous dinosaur provinciality. In: D. H. Tanke and K. Carpenter, eds. Mesozoic Vertebrate Life. Indiana University Press, Bloomington:310–28.

Looy, C. V., and C. L. Hotton. 2014. Spatiotemporal relationships among Late Pennsylvanian plant assemblages: paleobotanical evidence from the Markley Formation, West Texas, USA. Review of Paleobotany and Palynology 211:10–27.

Looy, C. V., R. A. Stevenson, T. B. van Hoof, and L. Mander. 2014. Evidence for coal forest refugia in the seasonally dry Pennsylvanian tropical lowlands of the Illinois Basin, USA. PeerJ 2:e630.

Loughney, K. M., and C. Badgley. 2017. Facies, environments, and fossil preservation in the Barstow Formation, Mojave Desert, California. Palaios 32:396–412.

2020. The influence of depositional environment and basin history on the taphonomy of mammalian assemblages from the Barstow Formation (Middle Miocene), California. Palaios 35:175–90.

Loughney, K. M., D. E. Fastovsky, and W. G. Parker. 2011. Vertebrate fossil preservation in blue paleosols from the Petrified Forest National Park, Arizona, with implications for vertebrate biostratigraphy in the Chinle Formation. Palaios 26:700–19.

Lyson, T. R., and N. R. Longrich. 2011. Spatial niche partitioning in dinosaurs from the latest Cretaceous (Maastrichtian) of North America. Proceedings of the Royal Society B 278:1158–64.

Mackey, S. D., and J. S. Bridge. 1995. Three-dimensional model of alluvial stratigraphy: theory and application. Journal of Sedimentary Research B65:7–31.

Martin, R. E. 1999. Taphonomy: A Process Approach. Cambridge University Press, Cambridge.

Martinsen, O. J., A. Ryseth, W. Helland-Hansen, H. Flesche, G. Torkildsen, and S. Idil. 1999. Stratigraphic base level and fluvial architecture: Ericson Sandstone (Campanian), Rock Springs Uplift, SW Wyoming, USA. Sedimentology 46:235–59.

Maxted, J. R., M. T. Barbour, J. Gerritsen, V. Poretti, N. Primrose, A. Silvia, D. Penrose, and R. Renfrow. 2000. Assessment framework for mid-Atlantic coastal plain streams using benthic macroinvertebrates. Journal of the North American Benthological Society 19:128–44.

McCabe, P. J. 1984. Depositional environments of coal and coal-bearing strata. In:R. A. Rahmani and R. M. Flores, eds. Sedimentology of Coal and Coal-bearing Sequences. International Association of Sedimentologists Special Publication 7:13–42.

McCabe, P. J., and J. T. Parrish. 1992. Tectonic and climatic controls on the distribution and quality of Cretaceous coals. In: P. J. McCabe and J. T. Parrish, eds. Controls on the Distribution and Quality of Cretaceous Coal. Geological Society of America Special Paper 267:1–15.

McCarthy, P. J., and A. G. Plint. 2013. A pedostratigraphic approach to non-marine sequence stratigraphy: a three-dimensional paleosol-landscape model from the Cretaceous (Cenomanian) Dunvegan Formation, Alberta and British Columbia, Canada. New Frontiers in Paleopedology and

Terrestrial Paleoclimatology, Society of Economic Paleontologists and Mineralogists Special Publication 104:159–77.

McCune, B., and J. B. Grace. 2002. Analysis of ecological communities. MjM Software Design. Gleneden Beach, Ore.

Meffe, G. K., and A. L. Sheldon. 1988. The influence of habitat structure on fish assemblage composition in blackwater streams. The American Midland Naturalist 120:225–40.

Miller, K. G., M. A. Kominz, J. V. Browning, et al. 2005. The Phanerozoic record of global sea-level change. Science 310:1293–8.

Morgenthal, T. L., K. Kellner, L. van Rensburg, T. S. Newby, J. P. A. van der Merwe. 2006. Vegetation and habitat types of the Umkhanyakude Node. South African Journal of Botany 72:1–10.

Myers, N., R. A. Mittermeier, C. G. Mittermeier, G. A. B. da Fonseca, and J. Kent. 2000. Biodiversity hotspots for conservation priorities. Nature 403:853–8.

Nawrot, R., D. Scarponi, M. Azzarone, et al. 2018. Stratigraphic signatures of mass extinctions: ecological and sedimentary determinants. Proceedings of the Royal Society of London B: Biological Sciences 285:20181191.

Neal, J., and V. Abreu. 2009. Sequence stratigraphy hierarchy and the accommodation succession method. Geology 37:779–82.

Nyberg, B., and J. A. Howell. 2015. Is the present the key to the past? A global characterization of modern sedimentary basins. Geology 43:643–6.

Ohmann, J. L., and T. A. Spies. 1998. Regional gradient analysis and spatial pattern of woody plant communities of Oregon forests. Ecological Monographs 68:151–82.

Olson, D. M., E. Dinerstein, E. D. Wikramanayake, et al. 2001. Terrestrial ecoregions of the world: a new map of life on Earth. Bioscience 51:933–8.

Owen, A., G. J. Nichols, A. J. Hartley, and G. S. Weissmann. 2015a. Vertical trends within the prograding Salt Wash distributive fluvial system, SW United States. Basin Research 29:64–80.

Owen, A., G. J. Nichols, A. J. Hartley, G. S. Weissmann, and L. A. Scuderi. 2015b. Quantification of a distributive fluvial system: the Salt Wash DFS of the Morrison Formation, SW U.S.A. Journal of Sedimentary Research 85:544–61.

Paller, M. H. 1994. Relationships between fish assemblage structure and stream order in South Carolina coastal plain streams. Transactions of the American Fisheries Society 123:150–61.

Patzkowsky, M. E., and S. M. Holland. 2007. Diversity partitioning of a Late Ordovician marine biotic invasion: controls on diversity in regional ecosystems. Paleobiology 33:295–309.

2012. Stratigraphic Paleobiology. University of Chicago Press, Chicago.

Peppe, D. J., D. L. Royer, B. Cariglino, et al. 2011. Sensitivity of leaf size and shape to climate: global patterns and paleoclimatic applications. New Phytologist 190:724–39.

Pfefferkorn, H. W. 1980. A note on the term "upland flora." Review of Palaeobotany and Palynology 30:157–8.

Pitman, W. C., III. 1978. Relationship between eustacy and stratigraphic sequences of passive margins. Geological Society of America Bulletin 89:1389–1403.

Pope, M. C., S. M. Holland, and M. E. Patzkowsky. 2009. The Cincinnati Arch: a stationary peripheral bulge during the Late Ordovician. Special Publication International Association of Sedimentologists 41:255–76.

Quartero, E. M., A. L. Leier, L. R. Bentley, and P. Glombick. 2015. Basin-scale stratigraphic architecture and potential Paleocene distributive fluvial systems of the Cordilleran Foreland Basin, Alberta, Canada. Sedimentary Geology 316:26–38.

Rahbek, C. 1995. The elevational gradient of species richness: a uniform pattern? Ecography 18:200–5.

Rahbek, C., M. K. Borregaard, R. K. Colwell, B. Dalsgaard et al. 2019. Humboldt's enigma: what causes global patterns of mountain biodiversity. Science 365:1108–13.

Retallack, G. 1984, Completeness of the rock and fossil record: some estimates using fossil soils. Paleobiology 10:59–78.

Rheinhardt, R. D., M. C. Rheinhardt, M. M. Brinson, and K. Faser. 1998. Forested wetlands of low order streams in the inner coastal plain of North Carolina, USA. Wetlands 18:365–78.

Rheinhardt, R., T. Wilder, H. Williams, C. Klimas, C. Noble. 2013. Variation in forest canopy composition of riparian networks from headwaters to large river floodplains in the southeast coastal plain, USA. Wetlands 33:1117–26.

Rogers, R. R. 1993. Systematic patterns of time-averaging in the terrestrial vertebrate record: a Cretaceous case study. Short Courses in Paleontology 6:228–49.

Rogers, R. R., and M. E. Brady. 2010. Origins of microfossil bonebeds: insights from the Upper Cretaceous Judith River Formation of north-central Montana. Paleobiology 36:80–112.

Rogers, R. R., and S. M. Kidwell. 2000. Associations of vertebrate skeletal concentrations and discontinuity surfaces in terrestrial and shallow marine records: a test in the Cretaceous of Montana. Journal of Geology 108:131–54.

2007. A conceptual framework for the genesis and analysis of vertebrate skeletal concentrations. In: R. R. Rogers, D. A. Eberth, and A. R. Fiorillo. Bonebeds: Genesis, Analysis, and Paleobiological Significance. University of Chicago Press, Chicago:1–63.

Rogers, R. R., D. A. Eberth, and A. R. Fiorillo. 2007. Bonebeds: Genesis, Analysis, and Paleobiological Significance. University of Chicago Press, Chicago.

Rogers, R. R., S. M. Kidwell, A. L. Deino, J. P. Mitchell, K. Nelson, and J. T. Thole. 2016. Age, correlation, and lithostratigraphic revision of the Upper Cretaceous (Campanian) Judith River Formation in its type area (north-central Montana), with a comparison of low- and high-accommodation alluvial records. Journal of Geology 124:99–135.

Rogers, R. R., M. T. Carrano, K. A. Curry Rogers, M. Perez, and A. K. Regan. 2017. Isotaphonomy in concept and practice: an exploration of vertebrate microfossil bonebeds in the Upper Cretaceous (Campanian) Judith River Formation, north-central Montana. Paleobiology 43:248–73.

Rosendahl, B. R. 1987. Architecture of continental rifts with special reference to East Africa. Annual Reviews of Earth and Planetary Science 15:445–503.

Ryer, T. A. 1983. Transgressive–regressive cycles and the occurrence of coal in some Upper Cretaceous strata of Utah. Geology 11:207–10.

1984. Transgressive–regressive cycles and the occurrence of coal in some Upper Cretaceous strata of Utah, U.S.A. International Association of Sedimentologists Special Publication 7:217–27.

Sadler, P. M. 2009. Models of time-averaging as a maturation process: how soon do sedimentary sections escape reworking? In: S. M. Kidwell and A K. Behrensmeyer, eds. Taphonomic Approaches to Time Resolution in Fossil Assemblages. Paleontological Society Short Courses in Paleontology 6:188–209.

Sambrook Smith, G. H., J. I. Best, P. J. Ashworth, C. R. Fielding, S. I. Goodbred, and E. W. Prokocki. 2010. Fluvial form in modern continental sedimentary basins: distributive fluvial systems: COMMENT. Geology 38:e230.

Sampson, S. D., and M. A. Loewen. 2005. *Tyrannosaurus rex* from the Upper Cretaceous (Maastrichtian) North Horn Formation of Utah: biogeographic and paleoecologic implications. Journal of Vertebrate Paleontology 25:469–72.

Scherer, C. M. S., K. Goldberg, and T. Bardola. 2015. Facies architecture and sequence stratigraphy of an early post-rift fluvial succession, Aptian

Barbalha Formation, Araripe Basin, northeastern Brazil. Sedimentary Geology 322:43–62.

Scott, A. C., and G. Rex. 1985. The formation and significance of Carboniferous coal balls. Philosophical Transactions of the Royal Society of London, Series B 311:123–37.

Shanley, K. W., and P. J. McCabe. 1994. Perspectives on the sequence stratigraphy of continental strata. American Association of Petroleum Geologists Bulletin 78:544–68.

1995. Sequence stratigraphy of Turonian–Santonian strata, Kaiparowits Plateau, southern Utah, U.S.A.: implications for regional correlation and foreland basin evolution. In: J. C. Van Wagoner and G. T. Bertram, eds. Sequence Stratigraphy of Foreland Basin Deposits: Outcrop and Subsurface Examples from the Cretaceous of North America. American Association of Petroleum Geologists Memoir 64:103–36.

Sharp, I. R., R. L. Gawthorpe, J. R. Underhill, and S. Gupta. 2000. Fault-propagation folding in extensional settings: examples of structural style and synrift sedimentary response from the Suez Rift, Sinai, Egypt. Geological Society of America Bulletin 112:1877–99.

Siewers, F., and T. L. Phillips. 2015. Petrography and microanalysis of Pennsylvanian coal-ball concretions (Herrin Coal, Illinois Basin, USA): bearing on fossil plant preservation and coal-ball origins. Sedimentary Geology 329:130–48.

Smith, A. B., A. S. Gale, and N. E. A. Monks. 2001. Sea-level change and rock-record bias in the Cretaceous: a problem for extinction and biodiversity studies. Paleobiology 27:241–53.

Smith, R. M. H. 1993. Vertebrate taphonomy of Late Permian floodplain deposits in the southwestern Karoo Basin of South Africa. Palaios 8:45–67.

Spicer, R. A. 1989. The formation and interpretation of plant fossil assemblages. Advances in Botanical Research 16:95–191.

1991. Plant taphonomic processes. In: P. A. Allison and D. E. G. Briggs, eds. Taphonomy: Releasing the Data Locked in the Fossil Record. Plenum Press, New York:71–113.

Swenson, J. B. 2005. Fluviodeltaic response to sea level perturbations: amplitude and timing of shoreline translation and coastal onlap. Journal of Geophysical Research 110:F03007.

Tabor, N. J., C. M. Romanchock, C. V. Looy, C. L. Hotton, W. A. DiMichele, and D. S. Chaney. 2013. Conservatism of Late Pennsylvanian vegetational patterns during short-term cyclic and long-term directional environmental change, western equatorial Pangea. In: A. Gąsiewicz and

M. Słowakiewicz, eds. Palaeozoic Climate Cycles: Their Evolutionary and Sedimentological Impact. Geological Society of London Special Publication 376:201–34.

Trendell, A. M., S. C. Atchley, and L. C. Nordt. 2013. Facies analysis of a probable large-fluvial-fan depositional system: the Upper Triassic Chinle Formation at Petrified Forest National Park, Arizona, U.S.A. Journal of Sedimentary Research 83:873–95.

Van Wagoner, J. C., H. W. Posamentier, R. M. Mitchum, et al. 1988. An overview of the fundamentals of sequence stratigraphy and key definitions. In: C. K. Wilgus, B. S. Hastings, C. G. S. C. Kendall, H. W. Posamentier, C. A. Ross, and J. C. Van Wagoner, eds. Sea-level Changes: An Integrated Approach. Society of Economic Paleontologists and Mineralogists, Tulsa, Oklahoma: 39–45.

Van Wagoner, J. C., R. M. Mitchum, K. M. Campion, and V. D. Rahmanian. 1990. Siliciclastic Sequence Stratigraphy in Well Logs, Cores, and Outcrops. AAPG Methods in Exploration Series, No. 7, American Association of Petroleum Geologists, Tulsa, Oklahoma.

Voller, V, R., and C. Paola. 2010. Can anomalous diffusion describe depositional fluvial profiles? Journal of Geophysical Research 115: F00A13.

von Humboldt, A., and A. Bonpland. 1805. Essai sur la géographie des plantes. Levrault, Schoell, and Company, Paris.

Wadsworth, J., R. Boyd, C. Diessel, D. Leckie, and B. Zaitlin. 2002. Stratigraphic style of coal and non-marine strata in a tectonically influenced intermediate accommodation setting: the Mannville Group of the Western Canadian Sedimentary Basin, south-central Alberta. Bulletin of Canadian Petroleum Geology 50:507–41.

Wadsworth, J., C. Diessel, and R. Boyd. 2010. The sequence stratigraphic significance of paralic coal and its use as an indicator of accommodation space in terrestrial sediments. In: K. T. Ratcliffe and B. A. Zaitlin, Application of Modern Stratigraphic Techniques: Theory and Case Histories. Society of Economic Paleontologists and Mineralogists Special Publication 94:201–19.

Wanas, H. A., E. Sallam, M. K. Zobaa, and X. Li. 2015. Mid-Eocene alluvial-lacustrine succession at Gebel El-Goza El-Hamra (Shabrawet area, NE Eastern Desert, Egypt): facies analysis, sequence stratigraphy and paleoclimatic implications. Sedimentary Geology 329:115–29.

Ward, P. D., D. R. Montgomery, and R. M. H. Smith. 2000. Altered river morphology in South Africa related to the Permian–Triassic extinction. Science 289:1740–3.

Webber, A. J., and B. R. Hunda. 2007. Quantitatively comparing morphological trends to environment in the fossil record (Cincinnatian series; Upper Ordovician). Evolution 61:1455–65.

Weissmann, G. S., A. J. Hartley, G. J. Nichols, et al. 2010. Fluvial form in modern continental sedimentary basins: distributive fluvial systems. Geology 38:39–42.

Weissmann, G. S., A. J. Hartley, L. A. Scuderi, et al. 2015. Fluvial geomorphic elements in modern sedimentary basins and their potential preservation in the rock record: a review. Geomorphology 250:187–219.

Whittaker, R. 1956. Vegetation of the Great Smoky Mountains. Ecological Monographs 26:1–80.

1960. Vegetation of the Siskiyou Mountains, Oregon and California. Ecological Monographs 30:279–338.

1970. Communities and Ecosystems. MacMillan, New York.

Wilf, P. 1997. When are leaves good thermometers? A new case for Leaf Margin Analysis. Paleobiology 23:373–90.

Wilf, P., K. C. Beard, K. S. Davies-Vollum, and J. W. Norejko. 1998. Portrait of a Late Paleocene (Early Clarkforkian) terrestrial ecosystem: Big Multi Quarry and associated strata, Washakie Basin, southwestern Wyoming. Palaios 13:514–32.

Williams, J. W., and S. T. Jackson. 2007. Novel climates, no-analog communities, and ecological surprises. Frontiers in Ecology and the Environment 5:475–82.

Winemiller, K. O., and M. A. Leslie. 1992. Fish assemblages across a complex, tropical freshwater/marine ecotone. Environmental Biology of Fishes 34:29–50.

Wing, S. L. 2005. Depositional environments of plant bearing sediments. Paleontological Society Special Publication 3:1–13.

Wing, S. L., and W. Q. DiMichele. 1995. Conflict between local and global changes in plant diversity through geological time. Palaios 10:551–64.

Withjack, M. O., R. W. Schlische, and P. E. Olsen. 2003. Rift-basin structure and its influence on sedimentary systems. Society of Economic Paleontologists and Mineralogists Special Publication 73:57–81.

Woinarski, J. C. Z., A. Fisher, and D. Milne. 1999. Distribution patterns of vertebrates in relation to an extensive rainfall gradient variation in soil texture in the tropical savannas of the Northern Territory, Australia. Journal of Tropical Ecology 15:381–98.

Wright, V. P., and S. B. Marriott. 1993. The sequence stratigraphy of fluvial depositional systems: the role of floodplain sediment storage. Sedimentary Geology 86:203–10.

Wyant, J. G., R. J. Alig, and W. A. Bechtold. 1991. Physiographic position, disturbance and species composition in North Carolina coastal plain forests. Forest Ecology and Management 41:1–19.

Zaitlin, B. A., M. J. Warren, D. Potocki, L. Rosenthal, and R. Boyd. 2002. Depositional styles in a low accommodation foreland setting: an example from the Basal Quartz (Lower Cretaceous), southern Alberta. Bulletin of Canadian Petroleum Geology 50:31–72.

Zuschin, M., M. Harzhauser, and O. Mandic. 2007. The stratigraphic and sedimentologic framework of fine-scale faunal replacements in the Middle Miocene of the Vienna Basin (Austria). Palaios 22:285–95.

Acknowledgments

We appreciate the helpful discussions with Catherine Badgley, Kay Behrensmeyer, Susan Kidwell, and Ray Rogers that have influenced our ideas about the stratigraphic paleobiology of nonmarine systems. We thank reviewers Robert Gastaldo and Erik Gulbranson and editor Julie Bartley for their helpful comments. We especially appreciate Robert Gastaldo's extensive comments on our manuscript and for rich discussions of the stratigraphic paleobiology of nonmarine systems. SMH also thanks students of the UGA Stratigraphy Lab – Anik Regan, Samantha Khatri, and Max Deckman – for challenging discussions.

Cambridge Elements $\equiv$

Elements of Paleontology

Editor-in-Chief
Colin D. Sumrall
University of Tennessee

About the Series
The Elements of Paleontology series is a publishing collaboration between the Paleontological Society and Cambridge University Press. The series covers the full spectrum of topics in paleontology and paleobiology, and related topics in the Earth and life sciences of interest to students and researchers of paleontology.

The Paleontological Society is an international nonprofit organization devoted exclusively to the science of paleontology: invertebrate and vertebrate paleontology, micropaleontology, and paleobotany. The Society's mission is to advance the study of the fossil record through scientific research, education, and advocacy. Its vision is to be a leading global advocate for understanding life's history and evolution. The Society has several membership categories, including regular, amateur/avocational, student, and retired. Members, representing some 40 countries, include professional paleontologists, academicians, science editors, Earth science teachers, museum specialists, undergraduate and graduate students, postdoctoral scholars, and amateur/avocational paleontologists.

Cambridge Elements ≡

Elements of Paleontology

Elements in the Series

These Elements are contributions to the Paleontological Short Course on *Pedagogy and Technology in the Modern Paleontology Classroom* (organized by Phoebe Cohen, Rowan Lockwood and Lisa Boush), convened at the Geological Society of America Annual Meeting in November 2018 (Indianapolis, Indiana USA).

Integrating Macrostrat and Rockd into Undergraduate Earth Science Teaching
Pheobe A. Cohen, Rowan Lockwood, and Shanan Peters

Student-Centered Teaching in Paleontology and Geoscience Classrooms
Robyn Mieko Dahl

Beyond Hands On: Incorporating Kinesthetic Learning in an Undergraduate Paleontology Class
David W. Goldsmith

Incorporating Research into Undergraduate Paleontology Courses: Or a Tale of 23,276 Mulinia
Patricia H. Kelley

Utilizing the Paleobiology Database to Provide Educational Opportunities for Undergraduates
Rowan Lockwood, Pheobe A. Cohen, Mark D. Uhen, and Katherine Ryker

Integrating Active Learning into Paleontology Classes
Alison N. Olcott

Dinosaurs: A Catalyst for Critical Thought
Darrin Pagnac

Confronting Prior Conceptions in Paleontology Courses
Margaret M. Yacobucci

The Neotoma Paleoecology Database: A Research Outreach Nexus
Simon J. Goring, Russell Graham, Shane Oeffler, Amy Myrbo, James S. Oliver, Carol Ormond, and John W. Williams

Equity, Culture, and Place in Teaching Paleontology: Student-Centered Pedagogy for Broadening Participation
Christy C. Visaggi

Understanding the Tripartite Approach to Bayesian Divergence Time Estimation
Rachel C. M. Warnock and April M. Wright

Computational Fluid Dynamics and its Applications in Echinoderm Palaeobiology
Imran A. Rahman

A full series listing is available at: www.cambridge.org/EPLY

Printed in the United States
By Bookmasters